THE SPACE STATION DECISION

➢

NEW SERIES IN NASA HISTORY

Steven J. Dick, *Series Editor*

RELATED BOOKS IN THE SERIES

Single Stage to Orbit: Politics, Space Technology, and the Quest for Reusable Rocketry
ANDREW J. BUTRICA

NASA and the Space Industry
JOAN LISA BROMBERG

Space Policy in the Twenty-First Century
EDITED BY W. HENRY LAMBRIGHT

Faster, Better, Cheaper: Low-Cost Innovation in the U.S. Space Program
HOWARD E. MCCURDY

High-Speed Dreams: NASA and the Technopolitics of Supersonic Transportation, 1945–1999
ERIK M. CONWAY

Robots in Space: Technology, Evolution, and Interplanetary Travel
ROGER D. LAUNIUS AND HOWARD E. MCCURDY

THE SPACE STATION DECISION

Incremental Politics and Technological Choice

Howard E. McCurdy

THE JOHNS HOPKINS UNIVERSITY PRESS
Baltimore

To Brook

Printed in the United States of America on acid-free paper

Johns Hopkins Paperback edition, 2007
2 4 6 8 9 7 5 3 1

The Johns Hopkins University Press
2715 North Charles Street
Baltimore, Maryland 21218-4363
www.press.jhu.edu

The Library of Congress has catalogued the hardcover edition of this book as follows:

McCurdy, Howard E.
The space station decision : incremental politics
and technological choice / Howard E. McCurdy,
p. cm. — (New series in NASA history)
Includes bibliographical references.
ISBN 0-8018-4004-X (alk. paper)
1. Space stations. 2. Astronautics and state—United States.
I. Title. II. Series.
TL797.M334 1990
353.0087'78—dc20 90-30831 CIP

ISBN 13: 978-0-8018-8749-9
ISBN 10: 0-8018-8749-6

This work relates to NASA Contract No. NASW-4067.

CONTENTS

PART IV

PREFACE

At first, when the pioneers of rocketry realized that people could actually venture into space, the stages for the exploration of this new frontier seemed fairly clear. Prophets of space exploration drew up plans for rocket planes that could carry people from the surface of the earth into orbit and back again. They described space stations moving in orbits around the planet to which people could fly. Utilizing those outposts on the edge of space, they dreamed of ways to study the earth and heavens and travel to the moon and nearby planets.

In 1961, just three years after the United States placed its first satellite in orbit, President John F. Kennedy gave the fledgling space program coherence and direction by establishing the goal of reaching the moon. When the United States reached that destination in 1969, the coherence disappeared. For the second and third decades of space exploration, no similar goal guided the American space program. Those decades were marked by hesitation and disagreement about where the space program should go. People who wanted to fulfill the vision of space exploration by starting work on a space station during those decades discovered that much of the political support for their orbital dream had dimmed. Some policy makers wanted to increase space spending; others wanted to reduce it. Some wanted to establish a base on the moon and send an expedition to Mars; others wanted to study the solar system with robots and probes. Many believed that humans were destined to live and work in space; others questioned the wisdom of sending people into a hostile environment where machines, they believed, could perform just as well at a fraction of the cost and risk.

The absence of long-range goals complicated the space station decision to an unusual degree. A space station is a facility that contributes to the accomplishment of other goals. Much like an alpine base camp or a frontier fort, a space station makes other expeditions possible. It is not an expedition in and of itself. Building a space station is neither as glamorous nor as exciting as mounting an expedition into space. Questions about what kind of space station the United States might need—or whether the country needed one at all—could have been resolved most easily by weighing the contribution of an orbiting outpost to the nation's long-range plans for future expeditions. Without such plans, however, this could not be done. The space station decision, made in a policy vacuum, was skewed in ways that would not have been tolerated when the space program began.

From the voyages of Columbus to the conquest of Antarctica, visionaries in search of patrons have turned to governments and the treasuries they control. In the case of the space station, the bearers of the vision joined the government and worked from within. They became civil servants in the National Aeronautics and Space Administration. Committed as they were to human flight and the exploration of space, they accepted almost without question the need for an orbital outpost circling the globe at least two hundred miles above the surface of the earth. They asked for the money to start one as part of the 1961 Apollo decision that dispatched Americans to the moon. They asked for the money to build one in tandem with the 1972 decision to build the space transportation system. And they asked again as the space shuttle test program neared an end in 1982. This book examines the effort to get the space station approved and the effect that the lack of political consensus about the space program had on it. It examines the way in which NASA officials, working through devices like a special Space Station Task Force, made plans for a space station in the absence of any politically approved plans for the long-range exploration of space.

President Kennedy's selection of the lunar landing as the guiding objective for the American space program determined the fate of the orbital station during the first decade in space. The eight-year deadline for landing humans on the moon ruled out the construction of a space station as an intermediate step in that special journey. In 1969, the year in which Americans first landed on the moon, NASA officials sought approval for a space station by presenting it as one of the key steps in a twenty-year exploration plan. In the wake of the Apollo program, this seemed like the technically rational thing to do—to maintain long-term goals as a guide for near-term projects.

The strategy failed. Neither the space station nor the long-term plan was approved. Sensing the limits imposed by the lack of consensus, NASA officials decided to pursue the steps in their long-term plan incrementally, one at a time. First they would build the space shuttle, then the space station. Where necessary, they would rely on the tactics of incremental politics to get these programs approved.

In making an incremental decision, policy makers begin from an established base (generally defined as what the agency did in the previous year) and direct their attention not to the overall goals of the program but merely to incremental changes within it. By moving forward or backward from an established base, policy makers can change public policy without making final decisions about the long-term direction they are taking.

Where disputes over goals are not resolved, policy differences must be settled through bargaining and maneuvering rather than the sort of ends-means analysis that characterizes comprehensive review.

The advantages of an incremental approach toward space policy seemed clear at the time. NASA officials and their allies could get new projects approved and maintain some momentum in the American space program even though policy makers did not agree on where that program should go. For that momentum, NASA officials paid a price. Incremental policies permit politicians to move forward or backward from the status quo as conditions change. Incrementalism promotes flexibility, a much prized commodity in the world of politics. Gains in flexibility, however, are paid for through losses in commitment. Politicians tinker with incremental policies, even after incremental policies are approved. For science and technology agencies like NASA, where individual projects can take a decade or more to complete, short-term tinkering conflicts with the long-term stability that career civil servants need to complete their work. NASA officials who embraced incremental politics thus found themselves implementing projects with far weaker long-term commitments than the agency had received for its voyage to the moon.

In 1989 President George Bush re-established a long-term plan for the exploration of space, one that reaffirmed the vision of human flight and planetary exploration. The Bush commitment re-emphasized the degree to which the tolerance for incremental policies has been shaped by the decision-making styles of different presidents. Since space policy carries heavy international ramifications, American presidents play a leading role in defining it. Different presidents have approached this responsibility in different ways. In making the decision to send Americans to the moon, President Kennedy brought about what is generally characterized as a "comprehensive" decision. He established an overriding purpose—a space spectacular at which the United States could beat the Soviet Union—and imposed that purpose upon his White House staff. In making the shuttle decision, President Richard Nixon did not impose a similar objective on warring factions within his administration. Lacking such a purpose, members of Nixon's White House staff found themselves hammering out short-term compromises between parties with unresolved points of view. President Ronald Reagan created a similar process in making the space station decision. He allowed the White House staff to debate the merits of a space station without establishing the purpose for which it would be built, stepping in to decide the fate of the orbital facility only when staffers could not agree. Given the absence of a presidentially imposed purpose, the shuttle and space station programs emerged from the White House in a much more compromised form than the Apollo program.

The way in which an initiative becomes official policy within the White House is just as important as the way in which a bill becomes a law in Congress. A president can shape the White House policy process by forcing broad objectives on warring factions and by the degree to which he delegates decision making to his staff. This book compares the ways in which three different presidents have shaped space policy, how they have allowed their White House staffs to assist them, and the way in which NASA officials have adjusted their proposal strategies to fit the different decision making styles.

As much as possible, I have tried to let the participants speak for themselves. Readers should know what people from the bureaucracy, the White House, and Capitol Hill were thinking when they argued the wisdom of establishing a permanent human presence in space and made the moves that carried the decision through the governmental system. I have used their words, taking them from interviews, speeches, reports, articles, and hearings. In order to preserve the narrative flow, I have departed sometimes from the precise order in which the words were spoken. Although the rough chronology of events has been observed, the reader will find that the book is organized around the strategies that the participants adopted and the moves they made, rather than the thin straight line of vernacular time. In space, after all, time is relative, especially as one looks back from a distance.

Howard E. McCurdy
Washington, D.C.
December 1989

ACKNOWLEDGMENTS

Completion of this book would not have been possible without the support of the NASA History Office. Many of the documents cited in this book were available in the archives of the NASA History Office when my research began; more were added as a result of NASA's multifaceted effort to document the history of the space station. Additional information on subjects covered in this book can be obtained from the NASA History Office or from the author. Transcripts of the most important interviews used in this book are also available in the History Office, including those conducted by the author. (In the notes, citations of interviews conducted by the author omit the author's name.)

Dr. Sylvia D. Fries, Director of the NASA History Office, provided constant support throughout the project through her commitment to an intellectually challenging history program. Lee D. Saegesser, Archivist for the NASA History Office, could find any fact about the thirty-year-old space program that I needed. His knowledge of the space program in general and the files that jam the History Office always amazed me. He also helped locate many of the photographs that grace this book. Dr. Adam Gruen began writing a book on the development of the space station shortly after I began my research on the space station decision. His book continues where mine ends, following the program through the design phase. His insight and humor, as well as his knowledge of space station details, assisted me greatly.

Members of NASA's Space Station Task Force, officials from the Office of the NASA Administrator, and people at the NASA field centers who participated in the space station decision were generous with their time and cooperative throughout the study. I especially want to thank Dr. Terence T. Finn, John D. Hodge, and Capt. Robert F. Freitag, members of the Space Station Task Force who launched and then helped run NASA's Office of Space Station, and Philip E. Culbertson, who directed the space station program in its formative stages. Their support and commitment made this project possible.

Dr. W. Henry Lambright of the Maxwell School at Syracuse University made extensive comments on the manuscript, as did many of the people mentioned above. In seeking points of comparison for the space station decision, I was guided by the histories of the Apollo and space shuttle decisions prepared by Dr. John M. Logsdon of the George Washington University. He in turn commented upon my work as I prepared to publish parts of it in the journal *Space Policy,* of which he is the North American editor.

Two graduate students at the American University assisted me in collecting the documents necessary to construct this story: Michael Bindner and Gregory Lubel. My colleagues in the School of Public Affairs at the American University encouraged me to undertake this study, gave advice, and allowed me the time to research and write the story.

Writing contemporary history is no small challenge, especially when Congress or a new administration threatens to undo the decision around which the story revolves. I had to make many judgments about the significance of events while they were still current, without the advantage of hindsight. The people whom I have acknowledged assisted me in this, sometimes by disagreeing with my point of view. The interpretations that I have placed upon those events are mine and mine alone.

Introduction

The Vision

James M. Beggs, the sixth administrator of the National Aeronautics and Space Administration, looked out at his audience and told a story. A professor at one of America's leading business schools, Beggs recalled, once gave his students a special assignment. The professor asked the students to summarize the most outstanding characteristics of American society. They should do so, the professor told them, in just two words.

The best answer came from a young woman. The two words she chose, Beggs retold, were "we advance."

"The key to that advancement," Beggs emphasized, "lies in the fact that we have a continuing urge to chart new paths and to explore the unknown."

Beggs looked out at the members of the Economic Club and Engineering Society of Detroit. He had flown into Detroit to deliver this speech on June 23, 1982, hoping to win new support for an old idea.

"That instinct drove Lewis and Clark to press across the uncharted continent," Beggs continued. "It guided Admirals Peary and Byrd to the icy wastes of the poles. It drove Lindbergh alone non-stop across the Atlantic and sustained twelve Americans as they walked on the moon."

"The compulsion to know the unknown built our nation," Beggs announced. "It spurred the creativity of our scientists and engineers so that today we lead and are indeed envied by the rest of the world in science and technology. It is clear that if we ever lose this urge to know the unknown, we would no longer be a great nation." Beggs hoped that the compulsion to explore the unknown would prompt the nation to take one more step into the future. He chose his words carefully. As NASA's sixth administrator, he was not the first to make this proposal.

"I believe that our next logical step," he said, "is to establish a permanent manned presence in low-earth orbit." "This can be done," he continued, "by developing a manned space station," a constantly occupied outpost some 250 miles above the security of the earth across the border defining outer space.[1]

Beggs called it "the next logical step," an essential part of NASA's overall vision for the conquest of space. To officials within the U.S.

government who had a different point of view, however, the proposal seemed anything but logical. Leaders of the President's Office of Management and Budget thought it a foolish idea, an extravagant waste of funds for a government with a deficit so large that public officials had to borrow more than 20 cents out of every dollar they spent. A majority of the members of the President's National Security Council opposed the idea, convinced that a space station would produce no military benefits while drawing funds away from more important priorities: rockets and space shuttles that could ensure the United States orbital access in a national emergency, reconnaissance satellites that could verify compliance with arms limitation treaties, and space-based weapons that could protect the nation against ballistic missile attacks. The President's science adviser counseled against the initiative, portraying the space station as an outdated idea made unnecessary by advances in technology and robotics. With the exception of the director of the Office of Management and Budget, David Stockman, all of these people favored a more ambitious space program. They did not, however, share the belief that the space station was the next logical step within it. There was, in 1982, when Jim Beggs made his space station speech, no political consensus about the future direction the American space program should take. Nor had there been any since 1969, after Americans first landed on the moon.

The President's science adviser, George Keyworth, had a much different vision of how to advance in space. Like Beggs, Keyworth dreamed about the discoveries waiting to be made beyond the surface of the earth. Unlike Beggs, Keyworth's dreams did not include a rotating cadre of astronauts garrisoned in space orbiting the earth every ninety minutes. The space station, Keyworth maintained, was an unnecessary diversion.

"Why don't we let the American people share the grand vision of the future of space?" he asked a conference of propulsion experts in June 1983. "Most advocates of a space station readily acknowledge that it is, in truth, only an intermediate step in a more ambitious long-range goal of exploring the nearby solar system."

"So if we're going to be asked to make an installment payment in the form of a space station, why don't we let the American people share the grand vision of the future of space—whether it's an orbital transfer vehicle to provide manned access to high geostationary orbits, a manned lunar station, or even manned exploration of Mars?"

"Any of the kinds of space stations being discussed," Keyworth pointed out, "would be expensive. Those of you who have heard me on this subject know that I question the return on that kind of investment, because we've yet to have explained just what would be done with that station once it's completed." Think of the alternatives, he urged his audience. "Without great cost or risk," Keyworth explained, the nation's space shuttle could take the place of a space station. A few engineering changes could turn the shuttle into "a flexible manned space platform for research and development," launching automated factories and staying in orbit to conduct experiments for up "to twenty or thirty days."

That would free up funds for more missions that would "bolster our science and technology base," he said, and "re-ignite the spirit of adventure that captured America in the past." Automated spacecraft could explore the planets at a fraction of the cost of human expeditions. Space telescopes could peer into corners of the universe where no people could expect to go. Robots had already sent back pictures of Jupiter and the rings of Saturn, dug up samples of dirt on Mars, and mapped Venus through its clouds. "Although they no longer dominate our space activities as they once did," Keyworth said, praising the automated satellites and planetary probes, "I confess that they still excite me the most."[2]

Keyworth wanted NASA to study the alternatives to a space station; Beggs believed that the logic of a space station was indisputable and viewed the call for alternatives as a diversionary tactic. "We've done these studies ten times before," said John Hodge, the leader of the Space Station Task Force that Beggs established in 1982 to help him make his case, "and it's not going to come up much different."[3]

Devotion to the space station idea flowed out of two specific commitments deeply rooted within the National Aeronautics and Space Administration. One was the seventy-year-old pursuit of human flight; the other, the meaning of exploration. Although NASA was a relatively young agency, created in 1958 after the Soviet Union orbited the first artificial satellite, the roots of the agency went back much further. The politicians who established NASA created it out of the research centers of the National Advisory Committee for Aeronautics, a government organization conceived at the turn of the century and established by Congress in 1915. The inspiration of two secretaries of the Smithsonian Institution, NACA came into being to conduct research on the machines that let humans fly. Over time, NACA became the world's premier organization for aeronautical improvement, attracting bright young engineers to its test and research facilities. Some of those engineers had already started work on the problems of space flight when NACA became NASA in 1959. Flight into space became an extension of flight into air, an essential part of the reason why those people went to work for the government in the first place.

During the first decade of human space flight, NASA engineers were obliged to abandon the ethic of the airplane and substitute what one astute observer called "the human cannonball approach" to outer space. Astronauts were placed on the tips of rockets originally designed to launch bombs. "All that was required was a test subject who could sit still and stand the strain," author Tom Wolfe observed. "Such a creature could scarcely be called a pilot. He would be a passenger with biosensors attached to his body. He would *splash down* in the ocean at the end of his ride—he wouldn't even land like a man." NASA didn't need pilots to ride in its capsules. "A monkey's gonna make the first flight," the test pilots at Edwards Air Force Base laughed. "It was true," Wolfe wrote. "Chimpanzees took the first suborbital flight (before Alan Shepard's) and the first orbital flight (before John Glenn's)."[4]

Between 1969 and 1972, NASA executives convinced President Richard Nixon to restore the airplane qua spaceplane to the realm of space flight. The space shuttle, which Nixon approved, revived the commitment to flight in an agency that by then had sped humans to the moon. "In a few years," Jim Beggs promised his audience in Detroit in June 1982, "a fleet of at least four shuttles will be flying routinely to and from space, ferrying into orbit commercial, scientific, and national security payloads." Pilots sat in a cockpit when the space shuttle Columbia roared into orbit on April 12, 1981, its maiden voyage. Astronauts John Young and Robert Crippen flew the space shuttle, performing critical maneuvers and landing the spacecraft like any other experimental aircraft at NASA's flight test center at Edwards Air Force Base in California.

Like many in NASA, Beggs saw the space station and the space shuttle as twin parts of the same mission. "The shuttle program originally was conceived to include a space station," Beggs observed. "More than a decade ago a total system was envisioned in which the shuttle would transport payloads routinely to such a station." In 1969, NASA executives proposed that the United States build the space shuttle and space station in tandem, simultaneously. From NASA's point of view, each facility reaffirmed the rationale for the other. To operate a space station, NASA needed the shuttle to ferry crew and supplies back and forth. To operate a space shuttle, NASA needed a destination in space to which it could fly. Without a space station, the space shuttle had no place to go, no destination except the landing strips back home. Airplanes flew from place to place, not just aimlessly around in the sky. Omitting the space station reduced the role of the shuttle to that of a space truck, ferrying satellites that in many cases could as easily be launched into orbit by machines alone. "We owe it to the nation," Beggs said, linking the space station to the space shuttle, "to make optimum use of this new capability of routine and reliable access to and from space. What better way to make full and complete use of the shuttle than to develop a manned space station in orbit tended and supplied by the Columbia and her sister ships?"[5]

None of this made sense unless one believed in the exploration ethic and the destiny of humans to move permanently into space, much as they had moved across the oceans to new frontiers centuries before. Without a space station, humans would be reduced to the role of occasional visitors, tending orbital machines better suited to withstand the rigors of space. Occasional visits might have technical merit, but they would not satisfy the purpose of exploration. Only a space station satisfied that purpose. A station, by definition, would give the nation "a stopping place" (as in railway station), "a pioneer settlement" (as in frontier station), and "an area of residence" to which government officials could be assigned (as in foreign station). "Even if it could be proved that functionally everything conceived of today could be done by robots," said Daniel Herman, chief engineer on the Space Station Task Force, "we would still want a permanent presence of man in space." The reason, he explained, "is that we think it is NASA's charter to essentially prepare for the exploration of space by man in the twenty-first century."[6]

Extension of the exploration ethic to outer space was nearly as old as the history of powered flight. Serious proposals for the territorial occupation of space began to appear shortly after the Wright brothers demonstrated that humans could power themselves above the ground on which they stood. Between 1911 and 1929, while Western nations busied themselves maintaining colonies on the surface of the earth, a handful of European scientists and engineers sketched out plans by which their nations could establish orbital bases above the land masses to which they had heretofore confined their territorial ambitions. Working from the mathematics of rocket propulsion and gravitation, the scientists explained the orbital mechanics and functions of outposts in space. Konstantin Tsiolkovskiy wrote from Russia, Hermann Oberth published his work in Germany, and Guido von Pirquet completed his plans from Austria-Hungary.[7]

As a young student growing up in the Germany of the 1920s, Wernher von Braun saw the advertisement for Oberth's book on space flight in an astronomy magazine. More than any other person, von Braun would be responsible for clarifying in the American mind the relationship between space stations and space exploration. Oberth's book, originally published in 1923, opened with the equation for the thrust of a rocket and became progressively more complex. In part three, however, the tone of the book changed. In relatively simple language, the author discussed the uses to which such a rocket could be put. The formulas disappeared; the author began to describe a station in space and how it could be used.

"If we should let a rocket of this size travel around the earth, it would constitute a sort of miniature moon." From such stations, Oberth promised, humans could study "every detail on the earth." They could construct large solar mirrors. They could conduct many types of physical and physiological experiments that "cannot be carried out on earth because of gravity. In space, telescopes of any size could be used, for the stars would not flicker." Radiation in wavelengths not visible from the surface of the earth, blocked by the atmosphere, could be studied from orbit.

"The observation station," the author continued, "could also serve as a fuel station," where hydrogen and oxygen could be stored in a liquid state and transferred to "a very powerful and long range vehicle." Such a vehicle would be "easily capable of making the trip to other bodies of the universe."[8]

Von Braun recalled the astonishment he had felt as a youngster mastering the equations of rocketry and digesting more books and articles on interplanetary flight. "It filled me with a romantic urge," he said. "Here was a task worth dedicating one's life to! Not just to stare through a telescope at the moon and the planets but to soar through the heavens and actually explore the mysterious universe! I knew how Columbus had felt."[9]

By the time von Braun reached the United States, following his work as a builder of missiles for the German military, the expression of national power through colonial occupation had passed its apogee. The German quest for *lebensraum* had been snuffed out by defeats in two world wars; the British had decided to grant independence to India, and the colonial system would soon crumble. Space, however, provided a new frontier into which the more developed nations could extend their technological and economic influence in the competition for international hegemony. It was also a frontier from which the developed nations could maintain visual control over foreign areas back on earth. Von Braun landed in America, where the concept of exploration as a means of national dominance was much more suited to conditions in the post–World War II period. Unlike the Europeans, who historically sought to extend their national values through the occupation of territory, the Americans preferred to advance their influence through technology, economics, and the occasional exportation of the American form of government. Space gave the U.S. government an arena into which it could extend its technology, expand its power, and demonstrate the superiority of the American system.

Von Braun never stated his space station vision in such overtly political terms. His exploration ethic nonetheless contained the same sort of enthusiasm for advancing Western values into foreign regions as had propelled Germany and her sister nations toward earlier efforts at territorial occupation. In 1952, while working on the Army rockets that would carry America's first artificial satellite into space, von Braun and five other writers presented their vision of the exploration of space. They published their ideas in *Collier's* magazine, a popular periodical with a wide circulation among Americans at that time. In his article, von Braun once again described the uses to which a permanently occupied space station could be put.

"Within the next 10 to 15 years," von Braun predicted, "the earth will have a new companion in the skies." Von Braun's "artificial moon," as he called it, would be "carried into space, piece by piece, by rocket ships." His 250-foot-wide, wheel-shaped satellite would rotate, creating a "synthetic gravity" equal to about one-third of the force felt by a person standing on the surface of the earth. Humans could use the wheel-shaped base as a launching platform for deep-space expeditions. "A short rocket blast, lasting barely two minutes," von Braun explained, could propel humans in a specially designed spacecraft away from the station and out toward the earth's nearest neighbor. "From this platform," he dreamed, "a trip to the moon itself will be just a step, as scientists reckon distance in space."

Beyond its value as a transportation node, von Braun concluded, the space station "will have many other functions." From their vantage point in space, "technicians in this space station—using specially designed, powerful telescopes attached to large optical screens, radarscopes and cameras—will keep under constant inspection every ocean, continent, country and city." The strategic importance of this capability would be enormous, von Braun explained. "Because of the telescopic eyes and cameras of the space station," he predicted, "it will be almost impossible for any nation to hide warlike preparations for any length of time." Other telescopes and cameras would point away from the earth, toward the heavens. Unlike the telescopes scanning the earth, these instruments would not rest on the rotating wheel. "There will also be a space observatory, a small structure some distance away from the main satellite, housing telescopic cameras," he said. "The space observatory will not be manned, for if it were, the movements of the operator would disturb the alignment." The space station, in von Braun's view, would consist of not only the rotating wheel but also co-orbiting platforms.

Remembering that he still worked for the U.S. Army, von Braun wrote that "there will also be another possible use for the space station—and a most terrifying one." It could be used as a platform from which to fight a nuclear war.

People like von Braun described the functions of a space station well before the first humans sped into space. With the exception of space manufacturing—the creation of ultrapure pharmaceutical products or metal alloys in the microgravity of space—all of the major functions of a space station had been identified by the time NASA was formed. To the people who shared in the vision, space station was not an idea that needed to be analyzed; it was a project waiting to be built. "Development of the space station is as inevitable as the rising of the sun," von Braun professed in a burst of historical determinism. "Man has already poked his nose into space and he is not likely to pull it back."[10] The vision within which the space station was contained would continue to guide NASA strategy through the decades ahead, even in the absence of a presidentially approved long-range plan.

In late 1959, scarcely one year after the creation of the new space agency, officials in NASA's Office of Program Planning and Evaluation proposed a long-range plan for the exploration of space. Project Mercury, they correctly predicted, would put an American in space in 1961 or 1962. NASA would be ready to use its Saturn rocket to attempt a flight around the moon "in about 1966–1967 and the establishment of a permanent near-earth space station in about 1968–1969." From there, a landing on the moon would take place "beyond 1970." At least in their initial planning, NASA executives saw the space station as an intermediate step toward the goal of landing Americans on the moon. More important, they did not see steps like the space station as ends in themselves. The space station would be a facility, the means by which other objectives could be accomplished. NASA's long-range objectives went well beyond the functions provided by a space station or even the challenge of landing on the moon. "The ultimate objective," NASA officials kept reminding each other, "is manned travel to and from other planets."[11]

There had been plenty of debate within NASA over the role of a space station in this overall program of exploration. Harry Goett, asked to form a planning committee on Manned Space Flight shortly after the creation of NASA, opposed the early construction of a large space station. Soon to be appointed head of NASA's space science center just north of Washington, D.C., Goett saw a large space station as an engineering task not sufficiently related to NASA's research goals. Even George Low, NASA's chief of Manned Space Flight, cautioned his colleagues about the space station project. Low wanted to go to the moon, and thought that NASA would have to eliminate "as many steps as possible" in order to get the lunar expedition approved and funded.[12] Officials at NASA's Langley Research Center, however, had already begun the technical studies necessary to construct an earth orbital space station. NASA executives were also preparing for a space station symposium with industry officials and other government agencies to discuss the problems of rendezvous and docking, the use of modules as building blocks, and the mechanics of power generation in outer space. All this took place prior to the 1960 presidential election, when NASA officials had yet to propel a single astronaut into space and had received no official approval for any space flight missions beyond the single-seat Mercury program.[13]

By mid-1960 such differences had been submerged beneath a single NASA line. NASA officials wanted approval for what they called "Project Apollo," a spacecraft capable of taking astronauts to the vicinity of the moon. "This spacecraft should ultimately be capable of manned circumlunar flight, as a logical intermediate step toward future goals of landing men on the moon and the planets," said George Low. In spite of his reservations about the space station, Low announced that "the design of the spacecraft should also be sufficiently flexible to permit its use as an earth-orbiting laboratory, as a necessary intermediate step toward the establishment of a permanent manned space station."

As Low astutely observed, "Official approval of this program has not been obtained."[14] NASA's long-range plan had no more status than an internal bureaucratic memorandum. It might reflect NASA's vision, but it was not the official policy of the U.S. government. To make it official, NASA executives had to carry their long-range plan into the political arena.

Within the government as a whole, the status of the space station in the overall long-term plan could be settled in one of two ways. It could be settled through negotiation, in which people with different visions compromised or bargained away their differences. Or it could be settled comprehensively, by weighing the uses of a space station against their contribution to the country's long-range goals. As pragmatic engineers and scientists, NASA officials were more comfortable with the latter. They preferred means-ends analysis to political bargaining. To conduct means-ends analysis, however, they needed a long-range goal. As of 1960, they had none.

PART I

I

The Race

(Spring 1961)

"Do we have a chance of beating the Soviets by putting a laboratory in space?," John F. Kennedy, President of the United States, asked in the spring of 1961. NASA executives had proposed the construction of a research laboratory orbiting the earth as an "intermediate step toward the establishment of a permanent space station," part of their overall but officially unapproved long-range plan.[1]

If the United States could not beat the Soviets in establishing a laboratory in space, Kennedy asked, could it beat them by soft-landing a robot on the moon or sending astronauts on a flight around the moon, two other elements in NASA's long-range plan? What about "a rocket to go to the moon and back with a man?" he asked. "Is there any other space program which promises dramatic results in which we could win?"[2]

President Kennedy asked these questions eight days after the Soviet Union used its SS-6 "A-1" rocket to launch cosmonaut Yuri A. Gagarin into space. On April 12, 1961, Gagarin became the first person to leave the surface of the globe in a rocket-powered spacecraft and go into orbit around the earth. He completed a single orbit in 108 minutes and landed in central Russia without any difficulties. The United States would not be able to put astronaut John Glenn into orbit for another ten months, and he would fly in a capsule less than one-third the mass of the Russian spaceship.

In seeking an American response to the Soviet triumph, Kennedy was prepared to make what many characterize as a comprehensive decision. He identified his principal objective: a "space program which promises dramatic results in which we could win." He asked his advisers to analyze the leading alternatives for achieving that goal. Once he had selected the best alternative, he was prepared to delegate to NASA the necessary authority to carry it out—the funds, the people, and the technical discretion to make the decisions that would determine how it would be done.

The *Washington Post* called Gagarin's flight "a psychological victory of the first magnitude for the Soviet Union." The Soviet Union had launched the first earth-orbiting satellite; now they had put the first human being in space. "The score is two to nothing," said one angry Congressman, "favor the Russians."[3] President Kennedy had succeeded Dwight D. Eisenhower in 1961 by presenting himself as a young, active, vigorous leader who would get America moving again. At a press conference following the Gagarin flight, Kennedy said, "We have to consider whether there is any program now, regardless of its cost, which offers us hope of being pioneers."[4]

One month earlier, in March 1961, Kennedy had been more cautious. The newly appointed NASA Administrator, James Webb, had come to Kennedy asking for money to start work on the next steps in NASA's long-range plan. Webb wanted to accelerate Wernher von Braun's large-rocket program and launch what NASA officials now called Project Apollo, named after the Greek god of music, prophecy, medicine, light, and progress. The von Braun group, which had started its large-rocket program under the direction of the U.S. Army, had joined NASA in 1960. As Webb presented it, the Apollo project would provide the United States with a spacecraft that could serve as a manned laboratory in earth orbit and take American astronauts on a flight around the moon.[5]

By early 1961, NASA's long-range plan called for America's manned space program to move down two tracks. Whereas the original long-range plan, prepared in late 1959, called for the establishment of a permanent near-earth space station prior to the exploration of the moon, the revised plan called for these milestones to occur simultaneously. By varying the configuration of the mission and propulsion modules attached to the three-person Apollo capsule, NASA officials could launch a manned orbiting laboratory and carry out circumlunar flights as early as 1967. This would be followed by construction of the permanent space station and the manned lunar landings, both "post-1970." To do this, Webb needed President Kennedy's approval for both the Apollo spacecraft and von Braun's large Saturn and Nova rocket programs.[6]

George Low, NASA's chief of Manned Space Flight, explained how the Apollo spacecraft would allow the United States to deploy an earth-orbiting laboratory. "For earth-orbital flight, the mission module can be considerably heavier than for circumlunar flight." The mission and propulsion modules were to be launched along with the Apollo capsule on top of von Braun's large rockets. In earth orbit, "this module can usefully serve as an earth-orbiting laboratory, with adequate capacity for scientific instrumentation and reasonably long lifetimes in orbit." It would serve as a mini–space station, a prototype on which NASA could conduct experiments in preparation for the "permanent" space station to be built after 1970. This was how NASA officials proposed to resolve the status of the space station in their long-range plan.[7]

At his meeting with President Kennedy, which took place before the Gagarin flight, Webb laid out the case for Project Apollo. "To make flights about the earth with multiple crews or trips to the vicinity of the moon, we must develop a new space vehicle and team it up with the Saturn booster," he argued. President Eisenhower, Webb explained, had "eliminated from his budget the preliminary design studies to begin this effort" and thus had "emasculated the ten year plan." Unless Kennedy reversed that decision, Webb told the President, "the Russians will, for the next five to ten years, beat us to every spectacular exploratory flight."[8]

Kennedy listened to Webb's arguments without making a decision. The next day, after meeting with his White House advisers, he decided to support the funding for von Braun's rocket program, but disapproved NASA's appeal for funds to begin work on the Apollo spacecraft. Kennedy was not ready to make a commitment to a new space program, especially when Project Mercury had not yet succeeded in putting an American into space. "Kennedy had not made up his mind," said one observer; "he felt he needed more time."[9]

Events did not give him more time. Nineteen days after President Kennedy put the Apollo program on the back burner, Yuri Gagarin took off on his orbital flight around the earth. Now it was Kennedy's turn to abandon caution. He called Webb back to the White House and redefined the problem. Kennedy did not adopt the NASA long-range plan that agency officials had been formulating since mid-1959. Instead, he quizzed Webb on the possibilities for beating the Russians. "Is there any place we can catch them? What can we do?" Kennedy demanded. "Tell me how to catch up."[10]

The Apollo program, as NASA Administrator James Webb had offered it to the President three weeks earlier, contained two distinct objectives. It would allow the United States to launch the prototype for a space station—a manned scientific laboratory. It would also allow Americans to send a spacecraft on a voyage around the moon without landing on it. Kennedy wanted to know whether or not the United States could accomplish either of those objectives ahead of the Russians. The answers he received were not encouraging.

"The Soviets now have a rocket capability for putting a multi-manned laboratory into space," responded Vice President Lyndon Baynes Johnson, who as chairman of the White House Space Council had been asked to prepare a memorandum answering Kennedy's questions. It took Johnson only eight days to reply. Never one to favor formal hearings or extensive analysis, Johnson pried his personal contacts in search of a new space strategy.[11]

Johnson put the questions to Wernher von Braun. "With their recent Venus shot," von Braun replied, "the Soviets demonstrated that they have a rocket at their disposal which can place 14,000 pounds of payload in orbit." Barely two weeks after Kennedy's inauguration as President, the Soviet Union had used a new "A2e" version of their SS-6 rocket to send an unmanned probe toward the planet Venus. "When one considers that our own one-man Mercury space capsule weighs only 3,900 pounds, it becomes readily apparent that the Soviet carrier rocket should be capable of launching *several* astronauts into orbit simultaneously. Such an enlarged multi-man capsule could be considered and could serve," von Braun concluded, "as a small laboratory in space."[12] The official NASA position was just as direct. "There is no chance," NASA executives told Vice President Johnson, "of beating the Soviets in putting a multi-manned laboratory in space."[13]

When von Braun's rocket team had been asked to describe their own plans for an orbiting space laboratory, they had replied that "the minimum payload capability required for a laboratory of this size is about 40,000 pounds." That was just an orbiting laboratory, not a space station. One of the smaller space stations under consideration by NASA at that time, with a radius sufficient to permit rotation and artificial gravity, weighed 171,000 pounds. How could the Russians put up a 14,000 pound capsule and call it a national research laboratory?[14] The answer was simple. A space station could be anything the government wanted it to be. Von Braun implicitly recognized this when he told the Vice President that an "enlarged multi-man capsule" could be called "a small laboratory in space" by its launchers—so long, one expects, as the cosmonauts did at least one laboratory experiment in it.[15] Given time, the United States might be able to launch a bigger and more impressive space station, but it would not be the first. From the White House point of view, that took the laboratory-in-space out of the running.

A similar problem affected NASA's chances to win approval for the second phase of its Apollo proposal, a voyage around the moon and back. "There is a chance to beat the Russians in accomplishing a manned circumnavigation of the moon," NASA officials told the Vice President.[16] But as von Braun explained, the chance was not very good. "The existing Soviet rocket could," von Braun predicted, "hurl a 4,000 to 5,000 pound capsule *around* the moon with ensuing re-entry into the earth atmosphere. This weight allowance must be considered marginal for a one-man round-the-moon voyage. Specifically, it would not suffice to provide the capsule and its occupant with a 'safe abort and return' capability, a feature which under NASA ground rules for pilot safety is considered mandatory for all manned space flight missions. One should not overlook the possibility, however," von Braun pondered, "that the Soviets may substantially facilitate their task by simply waiving this requirement." As with the orbiting laboratory, the Soviet leaders could improve their odds of winning that space spectacular by unilaterally changing the rules of the game.

To find an objective at which the United States could beat the Soviet Union, NASA officials had to look beyond the first few years in their long-range plan, leapfrogging missions in the process. "We have an excellent chance of beating the Soviets to the *first landing of a crew on the moon,*" von Braun predicted.[17] Other NASA officials endorsed the trip to the moon. "A possible target date," they said, "is 1967."[18] If elected to a second term, President Kennedy would be in the White House to greet the astronauts when they returned. Kennedy clearly favored the idea. At a press conference he announced, "If we can get to the moon before the Russians, we should."[19]

The old hands at NASA agreed. Hugh Dryden, who had served as the director of the National Advisory Committee for Aeronautics and was now the deputy director of NASA, had argued that the goal of the national space program "should be the development of manned satellites and the travel of man to the moon and nearby planets." George Low, NASA's director of Manned Space Flight, had urged the agency's planning committee to "adopt the lunar landing mission as its present long range objective." That was two years earlier, and now the President was prepared to hand the careerists one of their most desired objectives on an accelerated timetable.[20]

The accelerated timetable clearly worried NASA Administrator James Webb, a newcomer to the agency. An attorney with extensive government service, Webb had served as budget director for a previous President and as the Undersecretary of State. He had directed large corporations, both in the oil industry and on the board of directors of the McDonnell Aircraft Corporation. He understood how things could go wrong in a large-scale endeavor operating under tight deadlines.[21]

The lunar landing tore up NASA's timetable, eliminating and compressing a number of steps the agency had written into its long-range plan. The accelerated program would leave the agency without its laboratory in space and, at the conclusion of the program, deprive the agency of a permanent spacecraft available for voyages into deep space. "The most careful consideration must be given to the scientific and technological components of the total program," Webb wrote to the President's science adviser, Jerome Wiesner. "You and I," Webb said, must make sure "that this component of solid, and yet imaginative total scientific and technological value is built in."[22]

Wiesner remained unconvinced. Speaking of the lunar goal, he announced that the White House Advisory Committee on Science "would never accept this kind of expenditure on scientific grounds," and he said as much to President Kennedy.[23] Kennedy's new budget director, David Bell, made a similar case. "I was very skeptical," he said. "The question was, is it worth it? And what are we going to get out of it?" Nearly as much could be learned about space "by sending instruments out instead of men," he argued. "And the additional cost of sending men out wasn't going to teach you very much." He could have added that the cost of the program would be staggering, but Kennedy already knew that. "I did not argue at all that we couldn't afford it," Bell recalled, "because we could, did, can."[24]

Webb worried about public support. At that point in time, the United States had not yet put a single American astronaut into space, not even on a suborbital trajectory. The Mercury project was struggling. The first test flight of the Mercury-Atlas system, with the space capsule unoccupied, exploded mysteriously one minute into the launch. A second test capsule crashed into the ocean after the escape rocket on top of the spacecraft ignited prematurely. NASA had scheduled a medium-sized chimpanzee for a ride on that flight, but happily for the chimp, took the animal off the roster before the test. Two weeks later NASA attempted another launch. The Redstone rocket ignited, lifted four inches off the launch platform, shut down, and dropped back onto the pad, damaging the rocket and its tail fins in the process. The top of the Mercury capsule then blew off, deploying two reentry parachutes, which fluttered uselessly alongside the still-upright rocket.[25]

"We must keep the perspective," Webb cautioned, "that each flight is but one of the many milestones we must pass. Some will completely succeed in every respect, some partially, and some will fail."[26] Some, indeed, would fail. A good rocket, such as the Atlas-Centaur, could develop a reliability rate of 86 percent. It was considered exceptional for a rocket such as Thor-Delta to achieve a 93 percent reliability rate. Read in reverse, these figures could be terrifying. What they said was that even under the best of circumstances, the agency could expect one out of every fifteen rocket launches to fail in some way. Test pilots at Edwards Air Force Base who were eligible to ride in the single-seat Mercury capsule on top of the Redstone and Atlas rockets called it "spam in a can," referring to a form of pressed meat popular before refrigeration.[27]

President Kennedy was nonetheless determined to go ahead. He asked Vice President Johnson to answer the key questions and draft plans for an aggressive space program, one that would focus NASA's attention on the task of accomplishing a space spectacular before the Soviet Union did. In his final report to President Kennedy, Vice President Johnson settled all major policy questions affecting America's first full decade in space. He recommended that the United States "achieve the goal of landing a man on the moon and returning him to earth." The objective, he supposed, could be accomplished "before the end of this decade." His thirty-page report, which was actually drafted by a group led by

NASA Administrator James Webb and Defense Secretary Robert McNamara, scrubbed the 1967–1968 deadline NASA had originally proposed. Instead, Johnson's report recommended that the lunar landing take place "before the end of this decade." If an accident occurred (and it did), NASA would need the extra time.[28]

The officials who drafted the Vice President's final report laid out an ambitious space program to accompany the lunar goal: new rockets, the new Apollo spacecraft, accelerated funding for lunar robots to test the surface of the moon before the astronauts arrived, and communication, weather, and navigation satellites.

Political support for the program was so strong that Webb could be realistic about its cost. "Some people use a number as high as $40 billion to land a man on the moon," he remarked on the day after the Gagarin flight. "Others say half that amount." Webb and his assistants refined their estimate as Kennedy and Congress considered the goal. "It is much closer to $20 billion than $40 billion," said Webb's deputy, Hugh Dryden. Webb was not required to negotiate the cost of the program with White House budget officials as a condition for getting the program approved, nor was he required to be terribly specific about the exact costs of all the components. The report that he and Robert McNamara and their associates drafted for the Vice President passed over the question of cost casually. Building the rockets necessary for an accelerated space program would "involve the expenditure of billions of dollars over a period of a decade or more," the report warned, while exploration of the moon would "cost a great deal of money." By starting with an uncompromised cost estimate, NASA engineers had a fighting chance to accomplish the mission without any major cost overruns.[29]

The Webb-McNamara report did not answer one terribly important question. How would the United States actually reach the moon? As a design question, this was left up to NASA to decide. The Webb-McNamara report made no provision for a space station from which the astronauts could prepare for their journey and depart. The astronauts would, in mountaineering terms, have to make a dash for the summit without establishing a high-altitude base camp.[30] People like von Braun had always assumed that serious space exploration required a staging base suspended above the gravitational pull of the surface of the earth. The lunar objective provided the first serious test of that assumption. Lacking time to establish a space station, the United States would have to make what NASA engineers called a "direct ascent." The astronauts selected to make the voyage would have to take everything on their backs, including the fuel and the launch platform needed to blast off from the surface of the moon. This required an enormously large rocket, which the NASA engineers called Nova, after an exploding sun.

On the day before the Webb-McNamara group met to prepare its report, NASA successfully fired astronaut Alan B. Shepard atop a Redstone rocket through a suborbital trajectory 116 miles above the surface of the earth. The Redstone rocket, which performed well in spite of some rugged vibrations about ninety seconds into the launch, generated 78,000 pounds of thrust. The Nova rocket would have to be 154 times more powerful, the largest rocket ever made.[31] The von Braun rocket team, in a report issued one year earlier, had told NASA executives that they could not build a Nova rocket for a lunar expedition by the end of the decade. The Vice President's group nonetheless wrote into their report that "the advanced goal of [a] manned landing on the moon requires the development" of what they called a "very large launch vehicle (NOVA)," and they asked that funds available for the project be quadrupled.[32]

Von Braun pressed for an alternative approach. He called it "earth orbit rendezvous." Spectacular in concept, it required his team to launch a series of smaller Saturn rockets, which would be ready for flight by 1966 or 1967. A succession of Saturn rockets, as many as fifteen in one plan, would lift off from the east coast of Florida, forming a flotilla on the edge of interplanetary space. As they flew in formation, their crews would unpack and assemble the lunar spacecraft, attaching spheres of fuel from tankers rocketing up to join the orbiting fleet. Once prepared, the exploratory vessel would leave the mother ships on its voyage to the surface of the moon.[33]

It was a space station without the station. Although they would not gain a long-life base in orbit around the earth, the space pioneers would learn a great deal about rendezvous, refueling, and the construction of large structures in space. That could lay the groundwork for a space station in the next round of flights. Von Braun lobbied hard for his vision. By the end of the year, wrote one observer, "it was tacitly assumed at NASA headquarters that the mode would be earth-orbit rendezvous."[34]

The idea made sense to everyone but a small group of engineers and scientists at the agency's oldest aeronautical research center in southeastern Virginia. Located on a marshy stretch of the Back River near Hampton, Virginia, the workers at the old NACA Langley Research Center had been punching holes in conventional beliefs about aeronautics and astronautics for thirty-eight years before they were asked to form the nucleus of the newly created National Aeronautics and Space Administration. Langley was a research center, a national brain trust. Not only did its scientists and engineers run programs, they were also paid to think.

What they thought seemed preposterous. Instead of rendezvousing around the earth, they suggested, why not rendezvous around the moon?

"When first exposed to the proposal," said von Braun, "we were a bit skeptical." How could three astronauts possibly rendezvous and dock two tiny spacecraft 240,000 miles from the earth over the surface of a body they had never flown across before? Half of the time the two ships would be out of contact, shielded from mission control as they passed behind the dark side of the moon. If the two spacecraft failed to rendezvous, there was the awful possibility that one might become an orbiting tomb around the moon, a permanent monument to American incompetence in space.[35]

The Langley engineers pressed their case. "Do we want to get to the moon or not?" they asked. To dash to the summit, the astronauts would have to travel light. Lunar rendezvous let them travel in two tiny spacecraft built as small as humanly possible.[36] Von Braun understood. He shocked his staff by announcing that the Langley plan "offers the highest confidence factor of successful accomplishment within this decade" and abandoned his pursuit of earth-orbit rendezvous.[37]

In committing the United States to undertake the race to the moon, President Kennedy avoided a commitment to the overall plan for space exploration that the founders of NASA originally mapped out. He made a commitment to only one part of it, which for political and strategic reasons U.S. officials sought to pursue. Kennedy's commitment did not include a manned space flight laboratory or any expeditions beyond the lunar landing. In making the Apollo decision, President Kennedy in fact put the overall NASA vision on hold.

Kennedy's decision nonetheless gave the civilian space program the focus so many space advocates desired. That focus resolved the indeterminate status of the space station. None was needed, at least not for a quick sprint to the moon. NASA's instinct for pragmatic engineering and national goals took over as space station planning tumbled off the list of top priorities.

Years later, NASA officials and their allies would look back on the Apollo era as the "golden years" of space exploration, when the agency possessed a national commitment, an openly supportive public, and the funds to make its premier program work. As for the other elements of their vision, NASA would deal with those in the future. To get to the moon in just eight years, the Apollo astronauts had to travel light. NASA officials would worry about traveling heavy later, after they landed Americans on the moon.

2

One New Initiative

(January 5, 1972)

Little more than a decade later, Richard Milhous Nixon, the thirty-seventh President of the United States, sat at the desk in his California home, prepared to approve America's second major initiative in space. NASA Administrator James Fletcher and his deputy, George Low, had driven south from the small seaside town of San Clemente that morning to meet with Nixon and firm up NASA's largest program for the 1970s. The presidential retreat, a red-tile-roofed estate built in the classic Spanish-American hacienda style, sat on a 75-foot bluff on the southern California coast a few miles south of San Clemente. From the office in his home, President Nixon could turn his chair and look out across the Pacific ocean. It was a perfect California day, January 5, 1972. A light breeze blew ripples across the sea, made bright blue by the clear sky and the dry, 73° weather.

The prospects had seemed much different over that ocean twenty-nine months earlier, as Nixon stood at dawn on the bridge of the USS *Hornet,* 3,500 miles away, waiting to welcome the crew of Apollo 11 back from the first expedition to reach the surface of the moon. Low-lying clouds had moved in at dawn, leaving the sky a gun-metal gray. He and the 2,200 sailors standing on the decks of the recovery ship glimpsed the spacecraft as it flashed above a break in the clouds, riding a bright orange ball of fire to its earthly home.[1]

President Kennedy, the man who had sent Apollo 11 on its journey, was gone. During their battle for the presidency in 1960, Kennedy had joked that he carried a heavy burden. "Do you realize," he had observed, "I'm the only person between Nixon and the White House?"[2] Kennedy had defeated Nixon, but Nixon had come back to defeat the candidate of the Democratic party in 1968, which would have been Kennedy's last year as President had he not been assassinated. Not only did Nixon get to welcome the astronauts back from the moon but he also got to decide where the American space program went from there. NASA officials wanted to complete the steps in their still-unapproved long-range plan. Nixon had a different idea.

The lunar landing left the United States with an important legacy—not just the expedition to the moon or the picture of the earth as a distant blue-and-white globe. The voyage also gave the people of the United States an image of a successfully administered program. The Apollo expedition was widely viewed as a model government program. The program worked because its advocates had not been forced to compromise the concept in order to get it approved. It worked because program details were delegated to executive branch officials who knew how to carry them out. It worked because the politicians provided the funds necessary to make it work and because the political consensus underlying the program never disappeared.

NASA officials sought to maintain this legacy by winning official approval for ambitious new objectives to be accomplished during the second and third decades in space. Even though civilian space expenditures had started to wind down as the lunar landing approached, NASA officials continued to believe that the President and Congress would approve major new objectives. As the United States prepared for the actual landing on the moon, President Nixon asked his Vice President to establish a small group to prepare a "definitive recommendation on the direction which the U.S. space program should take in the post-Apollo period" along with "a coordinated program and budget proposal," an assignment not unlike the one President Kennedy had given Vice President Lyndon Johnson eight years earlier. While Johnson had taken eight days to issue his reply, Nixon's Vice President took 214.[3]

NASA officials presented their whole vision to the Space Task Group, as the Vice President's team was called. The Space Task Group in turn set out an agenda that would essentially ratify NASA's internal long-range plan. If the public was willing to maintain the level of effort that had been devoted to space activities during the Apollo era, the United States could dominate the peaceful exploration of space for the remainder of the century. Starting with a six-to-twelve person module, the nation could construct a space station in orbit around the earth large enough to eventually house fifty to one hundred men and women. Using the space station module, the United States could put a smaller station in orbit around the poles of the moon. From the lunar station, a base could be established on the surface of the moon.[4]

"In order to support the station," NASA officials reported, "efficient transportation to and from the earth is required."[5] The Vice President's group embraced a three-pronged space transportation system. A "chemically fueled shuttle" would operate between the surface of the earth and the space station; a "chemically fueled space tug" would move people and equipment around the earth's space station and from the lunar station to the surface of the moon; and a nuclear-powered spacecraft would move people and supplies from earth orbit to lunar orbit and beyond.

To "increase man's knowledge of the universe," the members of the Space Task Group wrote, the United States should additionally undertake an ambitious program of unmanned exploration: two space telescopes (one for visible light, the other for high-energy observation), unmanned probes to Venus and Mars, and robots to fly by all of the outer planets and across the top of the sun.[6]

"It is my individual feeling," Vice President Spiro Agnew said in summarizing the group's final recommendation, "that we should articulate a simple, ambitious, optimistic goal" to guide the overall course of this space exploration program. That goal should be "a manned flight to Mars by the end of this century," he announced.[7] "Mars is chosen because it is most earth-like, is in fairly close proximity to the Earth, and has the highest probability of supporting extraterrestrial life of all of the other planets in the solar system." Realistically, the Space Task Group predicted, the United States could be ready to go in 1986.[8]

NASA engineers were prepared, with as little as five years' notice, to launch the first component of the space station, using a two-stage derivative of the giant Saturn V rockets that were sending Americans to the moon. The rocket could lift into orbit a large cylinder, 33 feet in diameter. One five-story cylinder could house a crew of twelve. Eight cylinders, linked together like petals on a flower, would create the 100-person base.[9]

So confident were NASA executives of receiving approval for the space station that in the summer of 1969 they awarded two contracts to the North American Rockwell Corporation and the McDonnell Douglas Aeronautics Company to define the work that would need to be done to construct the facility. NASA officials then established a Space Station Task Group to prepare the agency to undertake the job and appointed as its field director Frank Borman, head of the first three-person crew to circumnavigate the moon.[10]

At the same time, NASA executives created a Space Shuttle Task Group to "develop material for a report" to the White House. The space shuttle would provide "low cost transportation" to the space station and "carry out other important space missions at greatly reduced operating costs," the announcement read.[11] Although they would be developed simultaneously, the space station was viewed within NASA as more important than the space shuttle. The space shuttle would "support the station," providing "airline-type operations" between the station and the surface of the earth. The shuttle would not carry the space station modules into orbit, as later planned.[12]

As the first American astronauts to return from the surface of the moon rendezvoused with the USS *Hornet* and President Nixon, NASA officials announced the award of the two space station study contracts. The two task groups began their work. Lightheaded in their moment of glory, NASA's space pioneers might be forgiven for thinking that the celebration of space exploration could go on forever.

Nixon knew that little public support remained for an ambitious space program. Public opinion polls confirmed the breadth of public disinterest. Since the race to the moon had begun, public support for "doing more" in space had risen above the 30 percent mark only once. The number of people who thought that the United States should "do less" consistently exceeded the number of people who supported new initiatives, softening only as the first lunar landing approached, then bulging out again. By mid-1970, after two successful landings on the moon, 50 percent of the public in poll after poll stated that the United States should "do less" in space. Only 20 percent thought that the United States should "do more." Every time NASA sent astronauts back to the moon, in spite of the extensive publicity generated by the event, opposition to the space program crystallized. It was as if NASA had been given a mission designed to make the agency self-destruct upon its completion.[13]

The Apollo program had been undertaken in response to a national emergency, a modern technological challenge as threatening as an old-fashioned war. Wernher von Braun had recognized the nature of the cold-war Soviet challenge in concluding his recommendations to Vice President Johnson in 1961. "I do not believe," von Braun warned, "that we can win this race unless we take at least some measures which thus far have been considered acceptable only in times of a national emergency."[14] With his decision to go to the moon, President Kennedy gave the space program an objective as uncompromising as President Franklin Roosevelt's demand for the unconditional surrender of Germany and Japan in World War II.

The race to the moon took eight years to win, the length of America's longest war. Nixon understood, as did most presidents who presided over postmobilization periods, how quickly the energy created by a national emergency could dissipate. With the goal accomplished, the public was anxious to bring the space forces home and get on with normal affairs. People like NASA Administrator Thomas Paine, who wanted the race to go on forever, needed to be brought back to earth.

Following the success of Apollo 11, the White House and Congress delivered five sharp blows to the American space program. Those blows dissipated the momentum built up by the lunar expedition. They sent out a signal that NASA and its allies could not expect an Apollo-style objective for the next decade in space and thereby sent NASA's long-range plan into a bureaucratic black hole.

Four days after the astronauts returned safely to the surface of the earth, the director of the Office of Management and Budget delivered a letter to NASA Administrator Thomas Paine setting out budget targets for the space program over the next eight years. The letter warned Paine that the budget office would work to freeze NASA's budget at $3.5 billion for the remainder of Nixon's term.[15]

Agency heads often received poison pen letters of this sort from the President's budget staff. Such letters could be ignored, at least temporarily, under the assumption that the budget director spoke for himself and not for the President. This was what Paine did, knowing that the Vice President would soon deliver a set of space goals to President Nixon that would require spending at twice the suggested rate.[16]

The second blow came from the President himself. Two days before Christmas 1969, Paine met with Nixon to appeal the budget director's recommendation for the upcoming fiscal year. If NASA was forced to hold spending below $4 billion, Paine told Nixon, it would have to halt production of von Braun's Saturn V rockets. This would devastate NASA's long-range plan. All of the space station modules, which also formed the building blocks for future missions like the lunar station, were to be launched from a derivative of the Saturn V. Without the Saturn rocket program, construction of the space station would have to wait for the development of a new launch vehicle, which might take a decade or more. A few days later, Paine received word that Nixon had rejected NASA's appeal.[17] Inspired by their victory on Paine's appeal, Nixon's budget director and other members of the White House staff proceeded to cut NASA's budget even deeper. In transmitting their decision to the NASA Administrator, members of the White House staff stated that "there is no commitment, implied or otherwise, for development starts for either the space station or the shuttle in FY 72," thus wielding the third blow.[18]

By then, Vice President Agnew's recommendations for the post-Apollo space program had been sitting on the President's desk for four months. NASA executives waited another two months before receiving Nixon's official reply. In a policy statement carefully drafted by the White House staff, the President delivered the fourth blow. Setting out his policy on space, Nixon announced that the space program should continue to make "steady and impressive progress." Indeed, he said, "space activities will be part of our lives for the rest of time." Nevertheless, he cautioned, "we must also realize that space expenditures must take their proper place within a rigorous system of national priorities. What we do in space from here on in must become a normal and regular part of our national life and must therefore be planned in conjunction with all of the other undertakings which are also important to us." No more "separate leaps," he said. No more "massive concentration of energy and will," not another "crash timetable." NASA, in other words, would have to get down on the ground and scratch for seed with all of the other government chickens.[19]

Congress delivered the fifth blow. NASA officials turned to their friends who sat on the space committees on Capitol Hill, hoping that they would restore some of the funds that the White House had cut from the manned space programs. The agency even trotted out Wernher von Braun. About to turn fifty-eight, von Braun moved from the rocket center in Huntsville, Alabama, to NASA headquarters in Washington, D.C., to lobby for more rockets, a space station, and a mission to Mars.

Von Braun arrived too late. Members of Congress read the same public opinion polls as the President. In a final gesture of support, members of the House Committee on Science and Astronautics convinced their colleagues in the full House to authorize a $300-million increase in the President's space flight budget. The increase disappeared, however, in the legislative battles that followed as members of Congress lined up to support President Nixon's conservative approach to space exploration.[20]

During the summer of 1970, as congressional support for an aggressive space program continued to evaporate, NASA Administrator Thomas Paine resigned. Von Braun followed him out of the agency two years later, taking a job as a traveling sales representative for a manufacturer of communications satellites. There was no political support for the Mars mission. Except for a small coterie of true believers, there was little support for a real space station, although NASA did retain funding to launch an orbital workshop built inside the upper stage of one of von Braun's unused Saturn V rockets.[21]

Within NASA, the implications of the events of 1969 and 1970 became clear. NASA officials could not win endorsement for a comprehensive long-range program of space exploration. They could not win approval for a new expedition in the Apollo mold. They could not even win endorsement for a space station and a space shuttle, two key steps leading toward the accomplishment of their long-term goals. "There is no way we can do that," said James Fletcher, who took over as NASA Administrator on April 27, 1971.[22]

Having failed to win presidential support for their internal long-range plan, NASA officials and their allies shifted their strategy. Rather than seek a comprehensive, Apollo-style commitment, they decided to pursue the steps in their plan one by one. Agency leaders would not abandon their vision, but would reach for it incrementally rather than comprehensively. An incremental approach begins with an established base made up of current agency programs, to which small changes are made. Progress occurs year by year through limited changes in the base, rather than through a comprehensive commitment to some far-off goal. Absolved of the need to set a specific long-range objective, parties to the policy making need not agree on where the government is going in order to participate. Incrementalism as a rule of thumb allows limited progress in the absence of political consensus about long-range goals, a consensus that NASA officials and their allies clearly did not enjoy in 1970.

Having moved toward an incremental strategy, NASA officials had to select the initiative with which they would begin. Though most people in NASA favored a space station, circumstances dictated that the agency begin with the space shuttle. The President's Science Advisory Committee (PSAC) had revived the argument about manned versus unmanned exploration that President Kennedy's lunar commitment had submerged. "There is insufficient information," a special PSAC panel reported in 1970, "upon which to judge" the relative advantages of "manned and unmanned—automated or remotely controlled—methods of space exploration." The science advisers recommended that Nixon not make a commitment to any new manned space endeavor until the government had explored the feasibility of reducing the cost of space operations. "Transportation of unmanned spacecraft to low Earth orbit typically costs about $1,000/pound," the committee observed. The members of the committee wanted to see that figure "drop below $100/lb" and predicted that "some form of multiple reusable space transportation system will indeed become the most attractive major new technological development for the decade."[23]

In a more practical vein, the budget cuts that NASA took in 1969 and 1970 had eliminated the only existing rocket program on which NASA could launch a space station. Without the Saturn V, the construction of the space station would have to wait incrementally for the completion of the space shuttle—assuming that NASA officials could get the shuttle approved.

The choice between the space station and the space shuttle was complicated by a third new start at the top of NASA's long-range plan. Beginning in the late 1970s, the five outer planets of the solar system would line up in such a fashion that robotic spacecraft could fly by all five in less than ten years using the gravitational pull of the larger orbs. Such a fortuitous alignment would not occur for another 179 years. A scientist by instinct and training, NASA Administrator James Fletcher desperately wanted to take the United States on an outer planet "Grand Tour." To do so, however, NASA had to launch the touring spacecraft between 1977 and 1979—the same time period set by the Space Task Group as the target for deploying the space transportation system and the first-phase space station.[24]

Fletcher had little difficulty making his choice. As soon as he arrived at NASA in the spring of 1971, he announced, "I don't want to hear any more about a space station, not while I am here." He decided to push ahead with the grand tour and with "a space transportation system that will save money and be cost effective in the long run." Even so, he "had no idea what the President would go for and what he wouldn't go for."[25]

As the centerpiece of their original proposal, NASA engineers wanted to build the ultimate spaceplane, a fully reusable shuttle. In their minds they mounted a spaceship with wings on the back of a rocket-powered airplane as big as an intercontinental jumbo jet. Pilots in the rocket-powered airplane would fly the spaceship out of the thick part of the atmosphere, where astronauts in the upper stage would light their engines for a trip into orbit as the pilots in the lower stage flew back to earth. Although the fully reusable shuttle was technologically sophisticated, it was not cost effective.[26] Analysts in the President's budget office told NASA officials to examine more closely the question of whether such a program could meet the standards of cost-benefit analysis.[27]

Searching for a low-cost configuration that would meet the cost-benefit criterion, NASA executives called in experts from a consulting firm, Mathematica, Inc., to study the economics of the shuttle. Experts in the firm replied that the fully reusable shuttle would be cost-effective only if NASA planned to send up a spaceship every eight days, an obviously unrealistic flight schedule. Mathematica recommended that NASA try another approach—a semi-reusable shuttle in which conventional rocket boosters would give the orbiter spacecraft the lift it needed to leave the ground.[28]

To prove that their plans were cost effective, NASA executives had to demonstrate that their shuttle had a lower overall operational cost than any other design that anyone else could think up. Engineers like to debate design. When confronted with the possibility of a new program, NASA engineers instinctively drew pictures to see how it might look. As Fletcher moved the shuttle program through the White House review process in the second half of 1971, the implications of the design problem sunk in. Unlike the scenario in 1961, when President Kennedy oversaw the decision to go to the moon, President Nixon in 1971 provided very little guidance for his White House staff. Nixon delegated the decision. His White House staff in turn sought to forge an agreement on the technical details of the program before taking the proposal back to the President. Faced with precious little agreement on the future of the space program, and even less on the uses of the shuttle, NASA officials found themselves negotiating the configuration of the space shuttle as a prerequisite for obtaining White House staff support.

All sorts of configurations floated by. As an alternative to the fully reusable shuttle, NASA officials proposed a thrust-assisted-orbiter shuttle, labeled TAOS. Officials in the Office of Management and Budget wanted Fletcher to consider a space glider without any engines which would soar back to earth after being launched into orbit on top of a Titan III missile. In another version, the budget analysts proposed a two-person space capsule mounted on top of a Titan rocket. They called it a "big Gemini" program, named after the two-person capsules NASA had used to conduct rendezvous experiments in preparation for the landing on the moon. Fletcher called it "a space stunt" and wanted no part of it. "Everybody," moaned one participant, "was a shuttle designer."[29]

As the end of the budget season approached in late 1971, Fletcher struggled to draw a compromise out of his meetings with the directors of the Office of Management and Budget, the science adviser, and the senior White House staff. Three of the officials negotiating on behalf of the White House told Fletcher that the President would accept a thrust-assisted-orbiter shuttle provided NASA kept it small, with a cargo bay only 30 feet long. NASA executives wanted a 60-foot cargo bay, large enough to carry military payloads and the modules that might be used to build a future space station.[30]

Officials from the Defense Department threw their weight behind the 60-foot alternative, resurrecting the old Apollo program argument that America needed a strong presence in space regardless of the cost and promising to use the big shuttle to launch and retrieve large military satellites. Other White House officials argued that only the 60-foot alternative was large enough to give the aerospace industry the boost it needed in an election year. Although he preferred the 60-foot-long configuration, Fletcher agreed shortly before his meeting with the President that NASA would recommend a 45-foot cargo bay if that sort of compromise was necessary to get the program approved.[31]

On January 5, 1972, from his office overlooking the Pacific Ocean, President Nixon indicated that he was ready. An aide brought Dr. Fletcher and George Low into the President's study. Nixon had a two-page statement on his desk announcing America's newest initiative in space. Thirty-five months had passed since Nixon had asked Vice President Agnew to convene the Space Task Group to identify America's next extraterrestrial goals. President Kennedy, by contrast, had taken only six weeks to make the lunar decision.

Fletcher brought to the meeting a model of the new launch vehicle, but in keeping with the uncertainty generated by the White House negotiations, it turned out to be the wrong one. Overruling his budget advisers, Nixon had decided to approve the 60-foot cargo bay. Although the scaled-down model showed the 60-foot bay, the boosters on the side of the orbiter were the wrong size. Tall and thin, the boosters were shaped so as to use liquid fuel. NASA officials and their contractors would eventually build solid-fuel boosters, largely to cut costs. Nixon was ready to approve the program anyway.[32]

Fletcher showed Nixon the model of the space shuttle. Nixon asked Fletcher and his deputy whether they considered the program to be a good investment. When they said yes, Nixon told them to stress the fact that the shuttle was not a "$7 billion toy." The President said that he wanted to involve other nations in the new initiative. He understood why foreign astronauts could not fly on the Apollo spacecraft to the moon, but it had disappointed him, and he wanted foreign nations to participate meaningfully in the shuttle program.[33]

The meeting lasted less than an hour. When it was over, Fletcher and Low drove back to the San Clemente Inn to explain the program to the press. Unlike Project Apollo, there was no speech to Congress, no major initiating event. One of the White House press aides handed out the President's two-page statement.[34] "The President met this morning for forty-five minutes with Dr. Fletcher, Administrator of NASA, and Dr. George Low, the Deputy Administrator," the aide said. "The meeting was to discuss the possibilities of future initiatives in the area of manned space exploration and as you now know, the decision was made to proceed with the space shuttle. The President's statement outlines the importance he attaches to this decision," the press aide astutely observed.[35]

"I have decided," the President's statement read, "that the United States should proceed at once with the development of an entirely new type of space transportation system." The new system, Nixon pledged, would "take the astronomical costs out of astronautics."[36] Fletcher promised that it would bring the cost of space launches down "by a factor of ten." Those claims would help sell the program to what promised to be a skeptical Congress, controlled by the opposition party.[37]

NASA's original cost estimates for a fully reusable shuttle ranged from $10 billion to $13 billion. In the fall of 1970, NASA executives established a cost goal of $8.3 billion for development of the shuttle program. The President's budget analysts insisted that NASA hold shuttle costs between $4.7 billion and $5.5 billion in order to win White House approval. In "winning" the thrust-assisted-orbiter shuttle with the 60-foot cargo bay, Fletcher received presidential support for a $5.5-billion program. Once the decision had left President Nixon's hands, budget officials forced Fletcher to accept a $5.15-billion target. As NASA engineers suspected, the agency could not develop the system for less than

$6.7 billion. Fletcher, however, did not have a strong enough political commitment from the President to demand the larger figure. Fletcher left that problem to his successor.[38]

By scuttling the space station and agreeing to a $5.15-billion space shuttle, Fletcher had hoped to leave NASA with sufficient funds to launch the Grand Tour. Supporters of projects like the Grand Tour had often argued that a less ambitious manned space flight program would leave more money for unmanned flight. "I didn't want to give up on that one too easily," Fletcher said.[39]

NASA got its answer on the Grand Tour from the director of the Office of Management and Budget, just before Fletcher and Low traveled to San Clemente. OMB would allow NASA to take the tour, but not to all five outer planets, just two. Fletcher decided that he could not win an appeal to the President. "I was willing to settle for Jupiter-Saturn," Fletcher announced, recognizing that the White House simply did not want to go ahead with two big programs at the same time.[40]

The shuttle decision gave NASA its political baptism. Unable to get their overall vision approved, NASA scientists and engineers plunged into the morass of incremental politics. They had to negotiate shuttle design details with the White House staff. They felt obliged to accept a technologically inferior program in order to win political support, and they had to engage in the game of bureaucratic politics, seeking outside support from groups like the military, who came to NASA's aid.

Had NASA officials been able to win approval for their long-range plan, they would have been able to present the shuttle program as they actually envisioned it. Within NASA, the space shuttle was viewed as a sophisticated technology for transporting people and equipment to an earth-orbiting space station, which in turn would support their exploration goals. Neither the space station nor the exploration goals had been approved, however, so NASA officials adopted a more utilitarian rationale. They turned to earth-bound arguments, in particular the cost effectiveness of the system for delivering payloads into orbit. They promised to make the shuttle cost effective when in fact their primary motivation for building it was not economic.

NASA officials viewed the shuttle decision as something of an anomaly, the result of having to negotiate program details without much consensus on long-range goals during a period of severe budgetary constraints. The further NASA got away from the shuttle decision, however, the more the Apollo program started to look like the anomaly. Apollo had received the sort of unbridled political support that NASA executives found increasingly difficult to get. Project Apollo had also left the United States, in spite of the success of the mission, without a space base, a space ship, or a heavy lift launch vehicle for the decades beyond.

NASA spent the 1970s designing and building the space transportation system. Division over the wisdom of NASA's unapproved but underlying vision deepened. In 1981, as the agency prepared the space shuttle Columbia for its first test flight in space, President Ronald Reagan selected James Beggs to become the sixth NASA Administrator. Beggs believed in the overall vision. He also believed that Reagan would back one major new initiative in space. Beggs prepared to announce his plans for the next step in NASA's incremental game plan at his confirmation hearings in the United States Senate.

3

Beggs

(June 17, 1981)

James Beggs strode up the steps of the Old Senate Office Building. When he had visited the capitol thirty-seven years earlier, as a cadet at the U.S. Naval Academy in nearby Annapolis, this had been the main office building on the Senate side of Capitol Hill. Ninety-six senators and their personal aides had fit more or less comfortably into one large building and an annex back then. Now the old guard had disappeared, the Senate bureaucracy had grown to fill three large office buildings, and Beggs arrived as a presidential nominee from corporate America rather than a young naval cadet from northern Texas.

Beggs entered the hearing room of the Senate Committee on Commerce, Science, and Transportation, a relic from the earlier days. Thirty-three gray marble pillars appeared to support an intricately cast plaster ceiling, lit in part by three pairs of tall windows that overlooked the Senate grounds. In the rear of the room, a white marble fireplace provided a mantel for a gold-framed mirror from which the committee chair might judge his appearance as he took his place at the opposite end of the room. The chairman sat at the head of a U-shaped table, elevated on a platform eight inches above the assembled crowd. Witnesses approached a felt-covered table at the foot of the platform, facing eighteen senatorial places arranged in an inquisitorial semicircle around the invited guest.

Committee members sat on chair cushions, their elbows resting comfortably on the committee table. Longer legs had been attached to the witness bench, so that a testifier not smart enough to select a chair with a cushion on it would find the table uncomfortably high. Even a witness as tall and composed as James Beggs would feel the table top approach his armpits as he sat down, inducing the unsettling sensation that he was shrinking in the presence of the senatorial council.

On the morning of June 17, 1981, Beggs had come to the committee room to be confirmed in his nomination by President Ronald Reagan as the sixth Administrator of the National Aeronautics and Space Administration. Most of the questions would be routine, dealing with the operation of America's civilian space program. The committee chairman also planned to ask Beggs for his views on America's next initiative in space, and Beggs was prepared to give his answer.

Beggs had worked for NASA before, during the glory days of the Apollo program. NASA's Administrator, James Webb, had asked Beggs to leave his career as an aerospace executive with the Westinghouse Electric Company and run NASA's advanced research program. It was an exciting job, overseeing the research centers in NASA and the work of some 17,000 scientists, engineers, and support personnel. Only forty-two years old, Beggs had reported directly to Webb. Webb had sent him off to testify on Capitol Hill and fly around the United States making speeches and meeting with NASA contractors. It had been his first government job, not counting seven years in the Navy.

Just ten months before the first Americans landed on the moon, Beggs had watched Webb leave NASA. He had watched the White House staff cut NASA's budget by $1 billion, a 20 percent reduction in just two years. He had watched scientists and engineers leave the agency, and in 1969 he had left too. He had taken a job in the Department of Transportation, the number two job in the department, the third-largest department in the federal government. Spending for transportation grew; spending for space shrank. Beggs spent four years helping the Secretary of Transportation run the department, testifying on Capitol Hill and fighting for agency programs with the White House staff.

NASA's budget kept declining. When Jim Fletcher had struck the deal with the budget examiners that allowed NASA to build the space shuttle, back in 1971, Fletcher had gotten a commitment that NASA's budget would be cut no more. Fletcher said that he had the commitment "in writing," signed by the Director of the Office of Management and Budget. The next year the President's budget examiners cut another one-half billion dollars from NASA's budget. Fletcher appealed to the new budget director, waiving the paper promise. The old budget director had resigned. The new director told Fletcher, "That was George's decision, not mine." For NASA, which had underpriced the shuttle program to begin with, budget cuts forced the agency to change the design and management of the shuttle program in ways that would have been unacceptable during the Apollo era.[1]

Beggs had watched NASA employees agonize over the cuts. When he received the call from the White House telling him that President Reagan wanted to appoint him NASA Administrator, Beggs replied, "I am interested." But before he accepted, he told White House aides, "I would want to know from the President what his intentions are."[2]

Ronald Reagan had won the 1980 election by promising to cut back the size of the federal government. As the new administration took office,

Reagan's budget director proposed that NASA take a $1-billion cut as part of the overall deficit reduction package.[3] If they wanted him to come in and "assassinate" the agency, Beggs told the White House aides, "then I don't think I'm your man. I can't close up the agency for you. My heart's too much with that program; I just couldn't do that."[4]

Beggs had a good job with the General Dynamics Corporation and a nice home on Woodoaks Trail in St. Louis, Missouri. After returning to the business world in 1973, he had worked his way into the position of Executive Vice President for Aerospace Programs at General Dynamics. To return to Washington, he would have to take a large cut in pay. He had not campaigned for the job as NASA Administrator. His résumé stated that he was seeking a job as "chief executive" of a "major company."[5]

The White House staff set up a meeting for Beggs with Reagan. "The meeting was relatively short," said Beggs, "as most meetings with the President are." Beggs had not met the President before, except casually when Reagan was still Governor of California. Beggs asked the President if he planned to wind down the American space program as part of his cutbacks in federal spending. "No," Beggs heard the President say, "he didn't. He didn't know a great deal about it, but he thought that it was one of those things that the federal government ought to do."

Beggs said that he would look into the space program and make some recommendations about it. Neither Beggs nor the President discussed anything specific. Beggs had no specific plans for the agency and, so far as he knew, neither did Reagan. The meeting took place in March 1981. The White House did not announce the Beggs nomination until late April. Beggs did not want to move to Washington until NASA had tried to fly the first space shuttle into orbit.[6]

The shuttle program was behind schedule, over budget, and undertested. Administrator James Fletcher had agreed to develop the new space transportation system with a budget of $5.15 billion. NASA employees could not meet this goal. The development program grew to $6.7 billion. Fletcher's successor, Robert Frosch, had to beg the White House and Congress for extra funds to make the shuttle fly. Unlike Project Apollo, the shuttle program enjoyed neither a strong political commitment nor an adequate flow of money. Frosch was treated like any other agency administrator over budget in a government long on promises and short on cash. Congress and the White House forced him to stretch the spending for the program out and to cut back on what were viewed as nonessential items, like the individual testing of shuttle engine components.[7]

Delay had followed delay. NASA had originally promised to fly the space shuttle into orbit in early 1978. That year, one of the three main engines that would propel the shuttle exploded on its test stand. The year before, the blades on the engine's high-pressure turbopump had failed.[8] As pilots took the space shuttle Columbia on a 17-minute evaluation flight around Edwards Air Force Base atop a Boeing 747, forty-one tiles fell off. Most were dummy tiles, not the real silica blockwork that would have to protect the orbiter and its crew from the heat of reentry, but at the launch site thousands of tiles had to be removed and rebonded. Pessimists felt that the tiles would peel apart like the teeth on a zipper when the shuttle roared into space, condemning the astronauts to a fiery death when they attempted to reenter the atmosphere.[9]

Instead of the liquid-fuel boosters that NASA engineers had originally proposed for the shuttle design, the Columbia would be helped into orbit by two solid-fuel rockets bolted onto the side of an external fuel tank. Solid-fuel rockets could not be shut off or throttled down. Once ignited, they stayed lit. The government put nuclear warheads on solid-fuel rockets, not people. No human beings had ever ridden solid-fuel rockets into space.

To supplement the power provided by the two solid-fuel boosters, astronauts in the space shuttle would have to rely on the shuttle's main engines for the extra thrust they needed to reach orbital velocity. To provide enough fuel for the main engines, NASA engineers designed an enormous, guppy-shaped external fuel tank, fastened it like an appendage to the underside of the orbiter, and pumped in 1.6 million pounds of highly volatile liquid hydrogen and oxygen. The external fuel tank had to hold together for 8 minutes of acceleration, then separate from the Columbia at 17,440 miles per hour, 68 miles above the earth.

Never before in the history of the space program had NASA asked its astronauts to pilot a rocket or a spacecraft into space on its maiden voyage. NASA engineers had always tested an unmanned vehicle first. Astronauts John Young and Robert Crippen would ride the shuttle into space on its very first orbital test flight. If the engines did not fail and the tiles did not peel off, then Crippen and Young would have to land the Columbia at Edwards Air Force Base on the first try. They would glide over the dry lake bed at 36,000 feet, make one sweeping turn, drop to the ground, and land. Very early on, NASA engineers had considered the possibility of putting two air-breathing jet engines on the orbiter, which would give it power to fly around in an emergency. The extra weight, however, did not seem worth the protection it provided. NASA engineers counted 748 different ways in which the two astronauts on the maiden voyage of the space shuttle Columbia could die.[10]

Beggs thought that the White House could wait until the Columbia flew before announcing his nomination. The acting administrator, he said, was doing a fine job and could answer all the questions put forth by the press about the details of the flight.[11]

The Columbia lifted off from the Kennedy Space Center on April 12, 1981, inaugurating the most dangerous flight since Americans had attempted to land on the moon. To the relief of NASA employees and contractors, who had labored to build the space transportation system for nearly ten years, the flight was a roaring success. A few tiles did come off, enough to "debond" the skin of the orbiter but not enough to burn a hole in the spacecraft. The cargo doors opened in space and then did not close properly, but the overlap was not enough to cause any problems during reentry. Other anomalies proved annoying but not dangerous. The controller for the cabin air temperature did not work properly. The zero-gravity toilet broke, a major inconvenience on a 52-hour flight. The crew returned safely, however, with their ship intact, giving the United States its first production-line access to space since NASA had been forced to cancel the Saturn V rocket program twelve years earlier.[12]

NASA executives still had to face a few cost problems on the shuttle program, but the production schedule was pretty much under control. NASA contractors were finishing work on a second orbiter, named Challenger, and had received the go-ahead to build two more. Three more test flights of the Columbia remained to be flown. On June 17, 1981, however, at his confirmation hearings, Beggs wanted to look ahead.[13]

He had talked to a lot of people about the future of the space program, inside NASA and out. One person played an especially important role. As Chief of Manned Space Flight in 1961, George M. Low had led the fight within NASA to select the lunar landing as the ultimate objective of the Apollo program. Ten years later, as NASA Deputy Administrator, he had traveled with Jim Fletcher to San Clemente to seal up the shuttle program as NASA's major space flight initiative for the 1970s. He had joined the NACA in 1949 as a twenty-three-year-old engineer and had left NASA in 1976 as its highest ranking civil servant. In 1980, as NASA began its third decade, the campaign staff for Ronald Reagan had asked the fifty-four-year-old Low to write a transition report on U.S. space policy for the president-elect.

Beggs remembered Low's transition report. "Typically on transition reports like that, usually the White House gathers them all up and burns them. And for good reason," Beggs observed. "A lot of them have some rather unfortunate recommendations that no new administration wants to live with."

"There were a lot of people," Beggs added, "who thought that they ought to burn this one before anybody read it. But I thought it was pretty well done."[14]

The lunar program, Low wrote in the 1980 transition report, had given the United States a position of leadership and preeminence in space. Lacking political consensus on where to take space policy after the moon landing, the United States had frittered away that lead. "NASA and the space program are without clear purpose and direction," Low reminded everyone. "It is recommended," his transition team wrote, "that a commitment to a viable space program be articulated by the President at a timely opportunity, such as the first flight of the shuttle in the spring of 1981." A viable space program could take many forms, the report said, "but it must have purpose and direction."

The transition team presented the incoming administration with three possible directions for the space program. NASA could transfer the shuttle program to another agency and focus on space science. It could "capitalize on the access afforded by the space shuttle" and emphasize science, exploration, and space applications. Or the President could make a commitment to a program "with a high challenge manned initiative."

"One possibility," Low's team wrote in explaining the third option, was "the establishment of a space operations center in low earth orbit." From such a space station, "it could be possible to construct and operate, perhaps by robotic techniques, large platforms in geosynchronous orbit for a multitude of applications and science programs. Such a program, or others like it, would open future space vistas, in the 1990s and beyond, that can only be imagined today: a permanently manned outpost on the moon or the human exploration of Mars."[15]

Beggs had no formal statement to read to the committee. After two months of preparation for his confirmation hearings, he knew what he wanted to say about the future of the space program and how to say it. He wanted to beef-up the space science area, which he thought the agency had neglected because of development problems with the shuttle. He thought that it was necessary to do a little reorganizing: the agency was getting old and he wanted to bring in some "new blood." He wanted to put a little more emphasis on aeronautics, a tough objective to sell in an age of astronautics. And he wanted the agency to undertake one new initiative. The next major goal in space, he had concluded after listening to all the advice, should be a permanently occupied space station.

"Little did we know how difficult it would be," Beggs observed. As required, he had cleared his testimony with the staff at the Office of Management and Budget. They had not told him to analyze more alternatives; they had not told him to be more circumspect. They would have plenty to say about the space station later, but for the moment they let Beggs make his pitch.[16]

Senator Harrison Schmitt called the committee to order, which on that day consisted of himself and two other senators. The junior senator from New Mexico, Schmitt had flown to the Taurus-Littrow Valley on the eastern edge of the Sea of Serenity nine years earlier as an astronaut on the last expedition to the moon.

Schmitt looked down at Beggs. "Today I am happy, on behalf of the Commerce Committee, to welcome Mr. James Beggs and Dr. Hans Mark who are nominated to be Administrator and Deputy Administrator of NASA, respectively."[17]

Alongside Beggs sat Hans Mark, a brilliant man with piercing eyes. He had emigrated to the United States from Germany as a boy with his parents. Earning his bachelor's degree in nuclear physics from the University of California at Berkeley and a doctorate from MIT, Mark had returned to the Berkeley campus as a research physicist. His colleagues made him a division leader at the University of California's famous radiation laboratory at Livermore and eventually chairman of the Nuclear Engineering Department at Berkeley. Under the Republicans, he had won appointment as Director of NASA's Ames Research Center, a high-tech national laboratory in the San Francisco Bay area. Then the Democrats had appointed him Secretary of the Air Force; there, he had made himself unpopular with the generals by pushing the space shuttle and manned space flight. He had wanted Beggs's job, but the Republicans had offered him the number two post in NASA instead.[18] Beggs had no say in Mark's appointment, even though Mark would be his deputy. The Reagan White House worked that way. Reagan's aides rarely allowed department heads to appoint their own deputies.

Schmitt asked the questions as the other members drifted away, the normal procedure for a noncontroversial confirmation. Schmitt asked about America's leadership in science and technology, about aeronautic research, about bringing new blood to NASA. It was 11:00 A.M. before he asked the new-initiative question Beggs was waiting for: "Setting aside the budgetary constraints that we are living with for the moment, and looking a year or two, fiscal year or two, ahead, what do you see is the next major step that is reasonable for our space program to undertake?" The space program, Schmitt added, had not had "an overall purpose since the statement by President Kennedy that we would go to the Moon." "It seems to me," Beggs replied, that "the next step is a space station."[19]

Beggs carefully understated the case. He had reached his conclusion by talking to a lot of people, not by ordering a formal analysis of the purpose of a space station and alternative ways of accomplishing that purpose. The people at NASA, moreover, remained deeply divided over what sort of space station they should build, and there were plenty of people in the White House and Congress and the science community who could be counted on to oppose the project.[20]

A space station, Beggs said vaguely, would "make a lot of other things possible in the future. The timing of that, I think, depends on how well we can bring the Shuttle along and get it operational." NASA officials would be conducting test flights of the space shuttle for at least another year. Given their incremental approach, they could not seriously advance a new program like the space station until the space shuttle had become an established program. Until then, none of the top policy people in Washington were likely to take talk about a space station seriously. NASA officials needed a very successful space shuttle upon which to raise the possibility of a space station.

"From what I can see," Beggs said, alluding to the first test flight of the shuttle Columbia, "we are off to a very good start." If the following flights went just as well, NASA could, "in the next few years, start thinking about a space station." Such a development, Beggs observed, would open up "all kinds of potential applications."[21]

The Senate approved Beggs's nomination by a voice vote on June 26, 1981, without any dissent. Vice President George Bush made the appointment official by swearing in Beggs at a White House ceremony on July 10. Beggs spent his first year as NASA Administrator working on projects other than the space station initiative. NASA had to get through the remaining space shuttle test program, which was due to end in the summer of 1982. Even so, Beggs had a number of opportunities to communicate NASA's desire for a space station to President Reagan and his White House aides. Beggs hoped that Reagan might approve the space station as a bold stroke, without the sort of protracted negotiations that had muddled the shuttle decision. Reagan, like President Kennedy before him, seemed genuinely entranced by the space program. NASA officials were pleased to have a true believer in the oval office again, one who might make an Apollo-type decision as President Kennedy had done two decades earlier.

Given all the other responsibilities he held, Beggs could not pursue the space station initiative alone. With the help of Hans Mark, he set up a small group in the office of the NASA Administrator to help him tell the space station story. If the opposition grew, Beggs would need all the help NASA could muster. On the other hand, the President might approve the space station as a bold stroke, relieving the group of the necessity of building a detailed justification for the new initiative.

4

The Team

(May 20, 1982)

"When I heard that Jim Beggs was coming back to NASA," John Hodge said in an easy-going way, his British accent still audible, "I [wrote] to him and said, 'Hey, if you've got a job over there, I'd like to come.'"[1]

Hodge missed his work with NASA. He had labored in the U.S. Department of Transportation for ten years, hanging on as funds for systems engineering ran down, and now he wanted to return to the agency that had brought him to the United States in the first place.

In NASA he had been a flight director, head of one of the teams at Mission Control in Houston that put the United States ahead in the race to the moon. He had been in charge on the day that Neil Armstrong and David Scott nearly died attempting to dock their Gemini 8 spacecraft to an Agena target vehicle, and had helped bring them home. After the Apollo capsule fire, which killed three astronauts on Launch Pad 34, NASA put Hodge in charge of the Apollo Crew Safety Review Board. Hodge marveled at how much of his career had been shaped by the need to figure out "how you can get virtue out of adversity."[2]

Born in England, Hodge grew up in a small town on the mouth of the River Thames east of London. His father, a teacher of brickworking at a vocational school, moved the family to Potter's Bar when World War II broke out.

Hodge finished high school as the war ended. "I was absolutely determined to be a biochemist," he said. "I got very high grades in chemistry and in math." But 90 percent of the already limited number of slots for first-year college students were reserved for soldiers returning from the war. "I couldn't get into biochemistry," he said, "but I could get into engineering."

"It's wonderful how your future is predicted by accidents," Hodge laughed. "Again, it's this whole question of making virtue out of adversity."

Hodge studied aeronautical engineering. "The aircraft industry wasn't doing very well in England," he remembered. "We moved to Canada in 1952 from England, my wife and I," to work for Avro Aircraft, building and testing the Avro Arrow, an supersonic interceptor fighter. He remembered his last day on the project. "We had stayed up all night, getting ready to fly the flight which would have got close to the world speed record of that day."

"That was on February 20, 1959," Hodge said, surprising himself that he would still remember the date, "at eleven o'clock in the morning."

"Over the P.A. system, the President of the corporation comes on and says, 'the government has just canceled the contract. We are no longer able to maintain the company. We are therefore closing down. Please everybody pack up your stuff and go home.'" Everybody lost their jobs. "Fourteen thousand people walked out of the building." It was known in aeronautic history as "black Friday."[3]

In the United States, NASA had awarded the contract to start construction of the Mercury capsule, the first American manned spacecraft. The agency could not get enough engineers. "Half a dozen of their people came up, headed by Bob Gilruth, who was running the Mercury program at that time," Hodge recalled. They arrived in Canada on a Friday. "They did all their interviews, we filled out form 52s . . . and on Sunday we got firm, fixed letters from the U.S. government offering us jobs and we didn't even have visas. On Monday they trooped [us] down to the U.S. embassy in Toronto, families and everything, kids all over the place. They gave us all the forms to fill out, we filled them all out and we had visas in four weeks." Hodge was amazed. "By the end of April there were thirty-five people down here working and I thought to myself, 'My God, if any country could work that well, particularly the civil service, and bend all its rules, then it has got to be something special.'"[4]

Hodge, a thirty-year-old engineer, helped set up the ground control systems for the Mercury, Gemini, and Apollo programs. When the Apollo program began to wind down, NASA put him to work on advanced planning, where he directed one of the space station studies run at the Houston center. But there were not going to be any advanced expeditions to follow the Apollo program. NASA cut its employment rolls nearly in half, and Hodge moved his career to the U.S. Department of Transportation.[5]

John Hodge and Jim Beggs left NASA for the Department of Transportation practically at the same time. Beggs took the number two post in the department, while Hodge directed advanced planning at the Transportation Systems Center in Cambridge, Massachusetts. They worked together on a number of high-tech transportation projects. Beggs eventually left to make a lot more money as a vice president of the General Dynamics Corporation. The Transportation Department began to deemphasize research and development.

"I was getting a little jaded," Hodge said, "seeing the demise of R&D in the Department of Transportation." So when Jim Beggs returned as Administrator of NASA, John Hodge wrote to him and asked to return. "I guess that it was in February of '82," Hodge said in his casual manner, when the executive officer to Jim Beggs "called up and said, 'Come over and talk to somebody.' "[6]

"Somebody" eventually turned out to be Philip Culbertson, a seventeen-year veteran of the space program. Culbertson had spent his entire government career at NASA headquarters, working through a variety of positions that combined an understanding of engineering with his knowledge of space policy. From his early days as a planner of new missions in the Apollo program office, he had worked his way into the office of the head of the space transportation system. When the White House told NASA to upgrade the oversight of the shuttle program in 1979, Culbertson got the assignment, working out of the office of the NASA Administrator. Once the space shuttle Columbia touched down, Culbertson started looking for a new assignment. Beggs and Hans Mark asked him to become what they called the Associate Deputy Administrator, their senior staff adviser. Culbertson agreed, asking that his assignments include the new space station proposal.[7]

Hodge was told to see Culbertson, "which I did, and then Phil says, 'Well, I don't know quite what the job is going to be but why don't you come over as sort of special assistant to me and start worrying about space station.' "[8] At the meeting, Culbertson decided that he would ask Beggs to put Hodge in charge of "a small office which we really didn't have until that time, put together an office under John reporting to me to pull together our story on space station." One month later Hodge was back working at NASA.[9]

When Hodge arrived, he found two other NASA employees working out of the Administrator's office on the space station initiative. A third person in another office was doing a lot of thinking about it too. Hodge looked over the backgrounds of the three people with whom he would be working. "These were unusual people," Hodge thought. "You join a task force," he remembered, "you leave your own organization. There's always that risk there might be no job when you get back." They had to be mavericks, willing to take risks, not bureaucrats, Hodge thought. They had to dare, he said, "to think and to discuss subjects which are beyond where we are and what is conventional."[10]

Terence Finn, at thirty-nine, was the youngest member of the group.

"Terry of course had this background in Congress," Hodge observed, a "tremendous knowledge of what makes Congress tick and how it works, a good solid feel for policy."[11] Finn had college degrees in political science, including a Ph.D. from Georgetown University, unusual in an agency where engineers and scientists got most of the promotions. His father was a senior partner in a big Wall Street law firm. Finn, who had grown up on the north shore of Long Island, had gone to a Quaker prep school and on to prestigious Williams College in northeast Massachusetts. "I was supposed to go to law school," Finn confessed. "I was supposed to be a New York attorney." But "what I wanted to do," he said, "was learn about the government."

"This is the strongest, greatest democracy in the history of the world, with a marvelous constitution," Finn said. "How it governs itself is eternally fascinating," he added. "I wanted to learn about it, and that meant doing it."[12]

So in the fall of 1965, instead of going to law school, Finn had enrolled as a doctoral student in political science at Georgetown University. His father helped him find a part-time job on the staff of a freshman senator from Maryland, Joseph Tydings. Finn worked his way up to the position of legislative assistant while struggling to complete his Ph.D. He served with the group that reorganized the committees in the House of Representatives. He managed an election campaign. He won an appointment to the staff of the Senate Budget Committee, where, among a dozen other assignments, he oversaw the budget of the National Aeronautics and Space Administration.

In 1978, thirteen years after he had arrived in Washington, Finn learned that NASA was looking for a new director for its Office of Legislative Affairs. Agency executives, in typical fashion, wanted to appoint a NASA insider. Finn applied and was turned down. But NASA's budget was in trouble. The space shuttle, which NASA had originally promised to fly that year, was grounded by delays and cost overruns. NASA needed someone who could work the budget process on Capitol Hill. Democrats controlled both houses of Congress, as well as the White House. Finn was a Democrat with very good ties on Capitol Hill. They called him back and offered him the job.

"All of a sudden we had to get a supplemental," a request for a special appropriation above the amount fixed in the annual budget. "NASA did not frequently have supplementals," Finn observed, but it needed one for the shuttle. "I tried very hard and we got the first supplemental through and Bob Allnutt [Finn's boss] writes me a note."

"'I got a copy of the bill,' he says. 'Terry, congratulations, you really did a good job. Guess what? We need another one.'"

"I thought he was joking," Finn said.

Finn laid out a second battle plan, worked to get the NASA Administrator involved, and helped marshal support from the U.S. Air Force, which had agreed to use the shuttle to launch its military satellites into space.

"It was a lot of meetings," Finn remembered, "and it was nurturing, going up there [to Capitol Hill] and talking to the staff. The supplementals broke the budget functional targets, so it meant going back to the budget committee staff, and I talked to those staff guys on every budget committee."

"So we got the second supplemental through. Then Allnutt tells me we need a third."

"I said, 'You've got to be kidding.'"

"The third time was a problem," Finn confessed. "Life got real serious around here."[13]

When the Reagan administration came in, tossing the Democratic party out of the White House, NASA had to change its Director of Legislative Affairs. For some reason, Finn said, "they didn't throw me out. They asked me to stay. Hans Mark in the summer of '81 asked me if I was interested in working on a task force that he and Jim Beggs were going to set up, to worry about a space station."[14]

Finn had been joined that summer by Daniel Herman. "Danny," John Hodge observed, "is just a first class engineer." Herman, Hodge learned, "had participated in a lot of the Code E programs—the science and applications programs—and particularly in programs that were nascent, that were beginning to happen, and what the process was associated with starting them."[15]

Herman had helped rescue NASA's planetary program. "I had a lot to do with selling Voyager," Herman recalled.[16] In the 1970s, scientists had wanted to send four robotic spacecraft past Jupiter, Saturn, Neptune, Uranus, and Pluto, a voyage that, because of a fortuitous alignment of the five outer planets, would not be possible for another 179 years. Jim Fletcher had dumped the space station in 1971 in part because he wanted to free-up money for what space scientists called "the Grand Tour." When support for the grand tour evaporated, along with support for the space station and the fully reusable shuttle, Herman and his associates lobbied for an ambitious alternative, the Voyager program. They proposed a program involving two spacecraft that could fly past Jupiter and Saturn and, if one of the Voyagers was designed well enough, on to Uranus and maybe Neptune.

"The very aspect that I could earn a living doing science on Jupiter and sending a spacecraft to Uranus or Neptune," Herman confessed, "I couldn't resist it."[17] Thinking about the future, Herman predicted that "at some point in time we are going to do a manned Mars mission." So why work on the space station initiative? "A manned Mars mission," Herman explained, his vision intact, "is a space station that is going to leave earth's orbit."[18]

Although an engineer, Herman had spent his whole NASA career on the science side of the space program. "I guess it's my father that led me in that direction," he remembered. Herman's father had immigrated to the United States from Poland in 1907. "He had no formal education, but he was probably one of the best educated men I've ever met. He was a vociferous reader. He was interested in science, never having had any formal education. He used to take me to the planetarium and the museum of science and industry when I was just six, seven, eight years old. I became very much interested at that early an age and sort of pursued a career in that direction ever since."

Herman's father owned a stationery store in New York. "I attended high school at the Bronx High School of Science and just naturally gravitated into an engineering profession," Herman recalled.

"During World War II, I was a radar technician in the Navy and actually I received my original engineering education from the U.S. Navy." At the end of his tour of duty, Herman returned to college under the G.I. bill, earning a degree in electrical engineering. He worked in the electronics and aerospace industry for twenty years, spending much of his time on NASA projects.[19]

When his employer, the Northrop Corporation, decided to sever its NASA ties, Herman moved over to the space agency itself. For eleven years, working out of NASA headquarters in Washington, D.C., he helped develop the agency's major planetary missions: Voyager, Pioneer Venus, Galileo, and the Venus radar mapping missions. He also developed very good ties to the science community.

In 1981, Hans Mark called Herman up from the science programs to work in the Administrator's office on space station. "The reputation that I had with Hans," Herman said, "was being able to sell impossible programs." The following year, at the age of fifty-five, Herman and Terry Finn welcomed John Hodge to the space station initiative.[20]

Of the three people Hodge would team up with, Capt. Robert F. Freitag was the only one that he had worked with before. "Bob Freitag," Hodge said, had "been around NASA for twenty odd years at that time and had twenty years in the Navy before and had been in the space business since the beginning of it.

"He helped set up the Cape and the Pacific Test Range and things of that kind and really was our resident knowledge base. Bob, it turned out, knew everybody, he really did. He knew all the history, had the most incredible set of files."

"You know, you talk about so and so, 'Do you remember that meeting we had, Chuck Matthews and Walt Williams, back in 1962,' and he says, 'Oh, yeah, I have that memo around,' and he'd pick it out and then he'd tell you the background behind it."[21]

Freitag had arrived in Washington forty years earlier, a twenty-year-old ensign freshly commissioned by the U.S. Navy at the start of World War II. The Navy was out to recruit all the aeronautical engineers it could find, commissioning them as they graduated from places like the University of Michigan, where Freitag worked his way through undergraduate school. The Navy gave him a salary and a quick graduate education at MIT. When Freitag completed his graduate studies at the top of his class, the Navy brought him to Washington, D.C., as a project engineer in its Bureau of Aeronautics.

Freitag remembered "the responsibility that people had thrust upon them in World War II. You never can imagine it in the present day." He was a lieutenant at that time and "just three layers away from the Congress." Reflecting on the present Navy, Freitag observed how "that same guy over there is probably three ranks higher, probably a commander, and he has ten or eleven to twelve layers of bureaucracy between him and Congress."[22]

The Navy sent Freitag to Europe to study German wind tunnels and guided missiles. He served on the operation that brought people like Wernher von Braun and the Peenemünde group to America. As a lieutenant commander, he was sent by the Navy to a remote peninsula along the Banana River on the northeast coast of Florida to lay out plans for the Joint Long Range Proving Ground at Cocoa Beach, which a decade later would become the nation's spaceport. "Saw it come into being," he said. "Saw the first launches."[23]

Back in Washington, Freitag struggled to convince the battleship admirals of the need for ballistic missiles launched from submarines parked underneath the surface of the sea. To win the support of the Chief of Naval Operations to initiate what would become the Polaris missile submarine program, Freitag developed the techniques he would use decades later in promoting the space station: build a clientele, talk to the White House staff, invite industries and government laboratories to submit technical proposals, and make, as one commentator observed, "literally hundreds of presentations."[24]

When the Deputy Secretary of Defense tried to kill both the Navy and the Army missile programs, Freitag forged an alliance with Wernher von Braun (who was developing the Army ballistic missile program) to save both. The defense secretary eventually endorsed the Navy's ballistic missile program and set up a Special Projects Office under the Chief of Naval Operations to develop it. The head of the Special Projects Office asked Freitag, who was then thirty-five years old, to become technical director for the Polaris program.[25]

Freitag turned it down. "I wanted to be in everything," Freitag confessed. He went out to California to help set up the Pacific Missile Range, the nation's West Coast launch facility. Back in Washington, he argued for a Navy space program and served on one of the committees that helped set up NASA. The Navy promoted Freitag to the rank of captain and made him its Director of Astronautics.[26]

"We had the navigational satellites, we had intelligence satellites, geodetic satellites, we had basic communication satellites," he said. "What we were doing is putting four or five satellites up on each launch vehicle and stringing 'em out like a Christmas tree."[27]

After twenty-three years with the Navy, Freitag came over to NASA. He moved through a succession of jobs: overseeing the rocket program, building up the field centers, repairing the damage done by Hurricane Camille, working out the international participation in the space transportation system. In 1973 NASA made him Deputy Director of Advanced Programs and told him to think up future missions for the space shuttle once it flew (some eight years later).

Growing up in central Michigan, the son of two generations of railroad engineers, Freitag had always dreamed of a future in aviation. "There was nothing I would rather do than fly," he remembered.[28] But the Navy turned down his application for flight school (bad eyes), and he got involved in guided missiles and Navy satellites instead. "It took a long, long time to accept that manned flight was as real to me," he said. "I wasn't comfortable that we could have a rocket like the Saturn V, with four billion parts in it, work effectively with only three minor failures. And I still don't believe that happened. Something else happened that made that work."[29]

Thinking up future programs was something that Freitag was very good at. There was widespread agreement within NASA at that time, Freitag said, that the space station was "the next logical step" beyond the space shuttle. "When the STS becomes operational," Freitag wrote in 1976, "we will apply our developmental resources to the second phase, the achievement of permanent occupancy in space. The major stepping stone of this second 'plateau' is the space station." In Freitag's view, it was inevitable. "It was just a matter of when."[30]

"It wasn't quite clear how we were going to structure the whole business of putting together the beginning of a program," Hodge said, remembering the first few days in 1982 when he had met with Finn, Herman, and Freitag. "The four of us started to work together." Hodge recalled. "Bob was enthusiastic about the idea of a task force and sort of said, 'The kind of thing we need to do is set it up with detailees, no permanent staff, the best people we could find in the agency,' to sort of get together and start to formulate plans for the program." Freitag recognized that they could not give Beggs the support he needed with just a small team of four in the Administrator's office. They would need a larger group—working out of headquarters but involving all the field centers—to put some detail on the space station concept and help Beggs promote the idea.[31]

Hodge agreed. At the end of March 1982, Freitag made a formal presentation to Philip Culbertson recommending that NASA create a Space Station Task Force. The task force, Freitag proposed, would "run 12–18 months, through initiation of phase B contract" (the point at which the program would move from just a concept within NASA to an approved initiative). Among the task force's responsibilities, Freitag listed "development of space station program plan" and "organize space station constituency." Point by point, Freitag outlined the functions the task force would need to perform.[32]

Culbertson also agreed that a task force was needed. "Everybody knew," he said, that the President "came in on an austerity, cut-down-the-deficit kind of a program." They also knew that they were going to have to have the support of a larger community outside NASA. "We at least had to have a situation in which we would not have everybody against us," Culbertson explained. "We also at that time felt that it probably should be a program which had the joint support of NASA and the Department of Defense." Culbertson believed that "the only way you can put together enough information on that basis, to make that kind of a rationale, would be to have something like a task force."[33]

On May 20, 1982, NASA Administrator James Beggs announced the establishment of a Space Station Task Force in a terse, three-sentence statement. "As a step toward a potential new initiative, NASA has established a Space Station Task Force under the direction of Mr. John D. Hodge," the statement read. "The Task Force is responsible for the development of the programmatic aspects of a space station as they evolve, including mission analysis, requirements definition and program management. In carrying out its functions, Mr. Hodge will report to Philip E. Culbertson, Associate Deputy Administrator, and will draw from the Space Station related activities of each of the NASA program offices and field centers." The task force, in other words, would work out of the Office of the Administrator at NASA headquarters, not out of a lesser office or field center.[34]

Freitag signed on as Deputy Director of the Task Force. When Finn and Herman joined, too, the number of members of the Space Station Task Force rose to four. "They were so interested in the job and the challenge," Hodge remembered. "Those first few months were absolutely marvelous."[35]

The members of the team knew that the Task Force would have to plan the space station concept: analyze its missions, define station requirements, and develop a management plan. The statement issued by Jim Beggs said that much. The announcement was also important for what it did not say. The Task Force leaders might have to get involved on the political side of the initiative, helping to build a constituency for the space station. The official statement made no reference to that, partly because NASA higher-ups were reluctant to admit that career civil servants could get involved on the politics side of policy making.

The degree to which Hodge and the members of his Space Station Task Force might find themselves working to build support for the space station depended upon President Reagan's approach toward the issue. If Reagan endorsed the initiative in July, as Jim Beggs and Hans Mark wanted him to, then the political responsibilities of the Task Force would shrink. Without such a commitment, however, the President would probably delegate the issue to the White House staff. In that case, Task Force leaders would have to mobilize support for the space station and represent NASA in the bargaining sessions that would surely follow.

President Reagan had agreed to make a speech at Edwards Air Force Base on July 4, 1982, following the fourth and final test flight of the space shuttle Columbia. Reagan aides planned to use the speech to announce the new White House space policy. Beggs and Mark thought that the occasion would provide the President with a marvelous opportunity to endorse the space station on its own merits, without elaborate staff review.

Beggs had been quite open about his intentions. At his confirmation hearings, he had promised to pursue the space station. He had told members of the press that he would recommend that the President publicly endorse a permanently manned space station. He had talked to the senior members of the President's staff, people like presidential counselor Edwin Meese, with whom he had regular contact, hoping to get such a commitment in the July 4 speech. He had told President Reagan about the space station—at a national security briefing on Soviet activities in space and at events involving the astronauts. While in Houston on presidential business, Reagan could not resist a last-minute change in plans that allowed him to visit Mission Control and sit at the CAPCOM console and talk to the crew of STS-2 in space. Three weeks later, on December 7, 1981, Reagan welcomed STS-2 astronauts Joe Engle and Richard Truly to the oval office for a White House ceremony commemorating their voyage. Beggs and Mark were on the scene both times, orchestrating events and talking to the presidential crew. On July 4, 1982, astronauts again drew Reagan to the landing at Edwards Air Force Base. Reagan liked American heroes, and the astronauts were among the best. As the astronauts returned from the last totally alien frontier, Reagan could reward those heroes with a permanent base to which they could fly in the future, if only he would say the words.[36]

5
Independence Day
(July 4, 1982)

Two Sikorsky S-61 helicopters flew in formation, one neatly aligned behind the other, as they crossed the Tehachapi Mountains and approached Edwards Air Force Base. The formation suggested that the Marine Corps helicopters did not carry ordinary people. White, roman-script letters spelled out THE UNITED STATES OF AMERICA on their green-and-white striped tails, revealing that one of the ships carried the President of the United States to the July 4, 1982, celebration on the high desert ground of southern California.

From his helicopter President Ronald Reagan could see the dry lake bed on which the Air Force had marked out landing strips for their experimental aircraft. Lines drawn with asphalt defined the different landing zones on the smooth clay surface. A concrete runway, three miles long, cut diagonally across the hard crust. Beyond the landing zones, near the hill the locals called Leuhman Ridge, the presidential party could see the crowd of people who had come to Edwards Air Force Base on America's Independence Day to watch the landing of the last test flight of the spaceship Columbia.[1]

Reagan carried with him a speech he would deliver that day setting forth a new American space policy. Supporters of the space station hoped that Reagan would use the occasion to identify America's next major initiative in space. It was widely agreed that any such decision would emanate from the White House. "The decision will have to be made by the President whether or not we ought to go ahead with the space station," said Edward Boland, chairman of the House Appropriations subcommittee that would have to approve the President's choice. "That will be a policy decision which will be made at the very highest level."[2]

Below the President's Marine 1 helicopter a long thin line of people pressed against the snow fence that ran along the edge of the landing zone. The morning sun outlined the cars, campers, vans, and recreational vehicles parked in orderly rows behind the fence, giving the area as seen from the air the appearance of a well-laid-out citrus grove. Cars and vans clogged the highways, as they had since the day before, when the crowd began to arrive. Now the traffic did not move, and the last of the five hundred thousand visitors to come to Edwards on the 206th anniversary of the approval of the Declaration of Independence left their vehicles and stood by the road and watched the President's helicopter fly by.

Two hundred miles above the California desert, Captain Thomas K. Mattingly maneuvered the spaceship Columbia into position for its descent to the landing zone. In the co-pilot's seat, fellow astronaut Henry W. Hartsfield scanned the instrument panel as Mattingly prepared to guide the Columbia back to earth.

Flight controllers on the ground told Mattingly and Hartsfield to prepare to land on runway 22. They would fly behind the crowd on Leuhman Ridge, make a 190° turn, and land on a concrete runway. Unlike the expansive lake bed, the 300-foot-wide concrete runway left little room for maneuvering. In this first descent of the orbiter to a conventional landing strip, touchdown would be similar to landing the spaceship at an airport. Still in orbit, the two astronauts passed over Edwards Air Force Base at 17,500 miles per hour. They would circle the globe one more time as the President's helicopter landed and the crowd counted the minutes down.

President Reagan's helicopter descended to the Delta taxiway in front of the row of hangars from which the test pilots flew. NASA's astronauts and test pilots fascinated Reagan. In their blue flight suits, they looked like heroes on a movie set. NASA had previously flown three test flights of the Columbia and had brought two of the astronaut crews to the White House to meet the President. The chance to talk to the crew of STS-2 from Mission Control had drawn Reagan to the Johnson Space Center outside Houston. Astronauts provided NASA executives with one of their primary routes of access to the President during his first eighteen months in office.[3]

Reagan walked down the steps from his helicopter to be greeted by Jim Beggs. NASA maintained a set of buildings and hangars on the northwest edge of the dry lake bed for testing spacecraft and high-performance airplanes. Known as the Dryden Flight Research Center, the facility was named for Hugh Dryden, who more than twenty-three years earlier had been one of the first NASA executives to call for the construction of a space station.[4]

Now that the test flights of the Columbia were over and the new space transportation system was ready, in Beggs's words, to start "going operational," Beggs wanted NASA to begin work on the space station. The landing of the Columbia was the milestone for which Beggs had been waiting.[5]

"By this time it was clear to me that the President favored the idea," Beggs said, referring to the space station. All that Beggs needed from the President was a public endorsement of the initiative. This was the time, Beggs believed, "to take the next step."[6]

Outside the spacecraft the earth turned dark. Flying backwards, with the tail of the Columbia pointed toward California, Mattingly and Hartsfield swept through the night of the day that would not come to California until that evening. As they passed over the Indian Ocean, computers in the spacecraft ignited two small orbital maneuvering engines mounted at the base of the spacecraft's vertical stabilizer. The engines fired for two minutes, just long enough to drop Columbia into an orbit 78 miles high, where the spacecraft would first sense the resistance of the atmosphere.[7]

On-board computers shut down the engines and turned the Columbia around so that its nose pointed forward. Mattingly could see blips of orange alongside the spacecraft, reflections in the night from the light of the control jets that were guiding their flight, bouncing off molecules of air, evidence that the ship had dropped into the upper layer of the earth's atmosphere. The night air around the spacecraft glowed a pinkish hue as friction from the atmosphere built up on the underside of the spacecraft.

The spacecraft banked to the left, pushing against the thickening atmosphere as the nose of the Columbia swung toward the north. The on-board computers slowed the orbiter like a skier descending a steep slope, banking the spacecraft to the left, then to the right, using the turns as a means of dissipating the tremendous energy created by its downhill speed.

The sun rose quickly under the nose of the spacecraft. Mattingly grasped the control stick and pulled it back a few degrees, taking control of the spaceship from the Columbia's on-board computers. He pulled the nose of the spacecraft up, then pushed it down so that the Columbia flew flatter than its normal angle of attack. Designed to test the spacecraft's landing range, the maneuver caused heat to build up rapidly on the pods protecting the orbital engines below the vertical tail. Mattingly then eased the nose back up to its normal reentry angle.

Halfway home from Hawaii and still 32 miles high, Mattingly and Hartsfield heard the voice of Mission Control astronaut Brewster Shaw as they emerged from their 16-minute communications blackout. Ground radar had locked onto the Columbia, and Shaw informed the crew that they were slightly off course. The on-board computers worked to correct the problem, changing the angle of the next set of banking turns by a few degrees. As the Columbia returned to its normal rate of descent, Mattingly took manual control of the spacecraft and began a second set of nose-up and nose-down maneuvers.

Beggs took Reagan to the two viewing stands NASA had constructed on the dry lake bed. The viewing stands were sturdy affairs, 12-foot-high platforms with open tops, dressed with American flags and the letters N-A-S-A. A plexiglass shield on one side blocked the wind. A plywood back hid their silhouettes from the distant crowd, a necessary precaution after Reagan had taken an assassin's bullet in the chest one year earlier and survived. The platforms were positioned alongside a blue-and-white fly-by tower, a structure on stilts which test pilots used to calibrate their airspeed instruments.[8]

Beggs introduced Reagan to the astronauts waiting on the viewing stand. Dressed in their blue flight suits, they were there to explain the landing of the Columbia to the President and Mrs. Reagan, who moved with them to the front edge of the viewing stand. "They hurried us up on the platform, because they said it was time to get up there, the shuttle was coming in," Reagan said. "They said it was on its approach."

On hearing that the Columbia was about to land, Reagan asked, "Where is it now?" From the platform, he and Mrs. Reagan had the best view of the landing strip. "Just over Honolulu," was the reply. Honolulu was five hours away by commercial jet. "I have to tell you," Reagan said later, "it was the biggest thrill I'd felt since hearing that Lindbergh had landed in France," an event which occurred when Reagan was sixteen years old. "The whole miracle," he said, "was brought home to me right then."[9]

The spaceship Columbia headed for the California coast like a guided missile, flying ten times the speed of sound. Sensors on the surface of the spacecraft detected the increase in air pressure on the outer skin as the orbiter dropped into denser atmosphere, signaling the computer to start to shut down the control jets and fly the craft with its ailerons. Dropping deeper into the atmosphere, the orbiter's elevators became effective, then the speedbrake on the tail, and finally the rudder itself. So accurate was the spacecraft's trajectory that a telescope on the California coast, programmed to point to an exact point in the sky, captured a picture of the Columbia as it passed through 160,000 feet, still flying over the Pacific Ocean.[10]

The Columbia passed over the California coast just north of the President's ranch. NASA's long-range cameras at the landing site produced a picture of the spacecraft as it crossed the Tehachapi Mountains, still 39 miles away. As the Columbia passed over the landing site at 36,000 feet, a sharp sonic boom brought the President's eyes up from the television monitors on the viewing stand to the spaceship turning in the sky.[11]

The spacecraft turned behind Leuhman Ridge and headed back toward the landing zone. Five separate navigation systems established the exact path of the Columbia and aligned it with the runway. Mattingly and Hartsfield scanned the instrument panel, verifying the trajectory of their final approach. "You just sit there and watch this thing stabilize itself," Mattingly said, praising the machine. "All the work is done by the flight control system."[12]

The on-board computers completed the turn. Just a few miles from the end of the runway, the Columbia still flew at 10,000 feet. The automatic landing system pushed the nose of the spacecraft down to a 20° angle and a descent rate of 10,000 feet per minute, twenty times as steep as that of a commercial jetliner. Captain Mattingly took over the controls at 2,500 feet, brought the Columbia back up to a shallow glideslope, dropped the wheels, and put the spacecraft softly on the runway.[13]

"Rolling to a stop," the CBS television news correspondent told his viewers. "The shuttle program is operational, we can say, can we not?"

"It certainly is," said the representative from Rockwell International, the manufacturer of the spacecraft.[14]

Mattingly and Hartsfield took 40 minutes to shut down the Columbia, run essential postflight tests, and wait to make sure that no explosive vapors had crept out of the orbiter's tanks or lines. The President and Mrs. Reagan, meanwhile, drove from their viewing stand by way of the base operations building to the concrete runway on which the Columbia sat.[15]

As support astronauts entered the cockpit to take over for the crew, Hartsfield started to get out of his seat, experiencing earth's gravity for the first time in seven days. "I sent the orders for my body to get up, but it did not get up. Basically, I felt like I weighed 800 pounds." He had the strength to get up, he said, but he also had something of "a sensory calibration problem."

"In zero-g," he explained, "your sensors get recalibrated, and to move limbs without gravity does not take a very big signal to the muscles to make it happen." Back on earth, he had to tell his body to try harder. "Once I got up," he said, "I proved to myself by jumping up and down and moving that it was not a strength problem at all."[16]

The two astronauts walked down the boarding ramp from the crew compartment and saluted the President, who stood at the foot of the ramp with Mrs. Reagan. "Reporting for operations," Mattingly said as he presented the Columbia to the President. Like a child with a new toy, the President inspected the spaceship, looking at the tiles that had absorbed the heat of reentry and so much of NASA's time. Mrs. Reagan reached up to touch one of the tiles on the nose-gear door. It was already cold.[17]

While the crew of the Columbia underwent their postflight physicals, Reagan prepared for the official part of his visit. After he was elected President, a number of people had told him that he ought to look at the space policy of the man he had unseated for his job, former president Jimmy Carter. The key words in Carter's space policy appeared on the first page of the old document.

"It is neither feasible nor necessary at this time to commit the United States to a high-challenge space engineering initiative," Carter's policy read.[18] At the ceremonies at the NASA facility, Reagan planned to announce a new policy. His staff would distribute a fact sheet officially outlining his new position. He would make a speech setting out the challenge.

Reagan's official space policy was a staff product. It had taken nearly a year to draft. Representatives from the White House, NASA, the State Department, the Defense Department, the Commerce Department, the Central Intelligence Agency, the Joint Chiefs of Staff, the Arms Control and Disarmament Agency, the Office of Science and Technology, the National Security Council, and the Office of Management and Budget had ground out the document. The President had directed the White House science adviser to coordinate the study, and the science adviser in turn had appointed the Assistant Director of the White House Office of Science and Technology to assemble an interagency working group.[19]

Out of this bureaucratic soup emerged an ambitious but relatively vague statement of principles. In place of Carter's policy of "no new initiatives," as space buffs liked to call it, President Reagan's official space policy committed the United States to "an aggressive, far-sighted space program." It targeted the fleet of space shuttles as "the primary launch system for both the United States national security and civil government missions," thereby committing NASA to an ambitious launch schedule. NASA's civilian space program, the document read, should aim "to preserve the United States preeminence" in the "exploitation and exploration of space." The policy statement did not, however, identify where that "far-sighted" space program should go.

As for the more immediate objective of a space station, the document blessed the efforts within NASA to "continue to explore the requirements, operational concepts, and technology associated with permanent space facilities." This was far less than NASA executives wanted. The policy statement did not even suggest, for a start, that the "permanent space facilities" should have people on board. It was, however, as much as the NASA representatives could pull out of the interagency group, given the level of disagreement still remaining over the future direction of the U.S. space program.[20]

The return of the astronauts provided President Reagan with an opportunity to step above the disagreements within his staff and initiate a new stage in the space program. Reagan could simply state his support for the space station to the assembled crowd of true believers. If the President was willing to deal with the initiative on a political rather than a bureaucratic level, he could settle the whole matter quickly, as President Kennedy had done when he took seven days to decide that the United States should race the Soviets to the moon.

NASA had certainly set the stage for such an announcement. On the grounds of the test flight facility, NASA officials had assembled their entire fleet of shuttle spacecraft. The spacecraft Challenger sat on the back of a Boeing 747 jumbo jet next to the runway on which the Columbia had just landed. In front of their Aircraft Construction and Modification Hangar, NASA officials had parked the Enterprise, the test vehicle used for suborbital landing trials. Underneath the wing of the Enterprise, they had built a platform on which the President could stand with the astronauts and deliver his speech. They had filled the taxiway in front of the Enterprise with 50,000 screaming space enthusiasts, special guests of NASA, all waving small American flags and standing so tightly packed that some had to sit on top of friendly shoulders to get a better view. As the President walked onto the platform, the Challenger would move to the Delta taxiway and out onto runway 23. As the President concluded his speech, the jet would roll down runway 23 and take off directly over the President's stand, banking to the right to give Ronald Reagan a clear view of the newest addition to the fleet of orbiters, the ships that could deliver the space station piece by piece into orbit around the earth.[21]

In addition, NASA executives had written "suggested remarks" for the President to deliver as the event took place. At the critical point in the speech, Reagan planned to proclaim the test program of the Space Shuttle Columbia successfully concluded. Hans Mark, NASA's deputy administrator, had provided the White House staff with words to follow. With the test program over and the shuttle ready to "conduct her first operational mission," it was time to move on to the next stage. And so, Mark and his colleagues hoped, the President would say that he was directing NASA "to initiate the first step toward the creation of a permanently manned American space station."[22]

The President's closest aides were involved, Beggs said, explaining why he thought Reagan might announce the space station initiative in the speech. Edwin Meese, one of the aides, had electrified the agency four months earlier by announcing on a nationally televised news and talk show that "we are supportive" of the space station program "and at the appropriate time I am sure there will be money in the budget."[23]

Beggs had followed up with Meese. "As we have discussed," Beggs wrote to Meese in May, experts "who are familiar with the technology" underlying the exploration of space "are in general agreement that the next major step should be a permanent manned space station in earth orbit."

"The President's potential presence at the Shuttle landing at Edwards Air Force Base," Beggs added, "provides an opportune time for announcing his plans. I strongly believe that these plans should embrace the development of a permanent manned space station." Beggs had followed up the letter five weeks later with the speech to the Economic Club and Engineering Society of Detroit kicking off the space station campaign.[24]

Such an announcement would allow NASA to circumvent the White House bureaucracy that had ground the agency down with the shuttle decision in 1971 and the official space policy the President was about to announce. Referring to such an effort by an agency head to circumvent the formal White House review process, one former presidential chief of staff noted that they used to call it the "oh, by the way" decision.

"That would happen when somebody was cleared to go into the oval office, a cabinet member, to talk about a subject," he said. "The President was prepared and they had their discussion. Then as the cabinet member was leaving he would turn around and say, 'Oh, by the way, Mr. President,' bring up a totally unrelated subject, get a decision on it, and run with it."[25]

"The most important thing that the chief of staff can do," another presidential aide said, "is to do everything within your power to see that the President is fully briefed before he does make a decision. It's very important for the 'Oh, by the way' decision not to get made, because nine times out of ten the 'Oh, by the way' comment or recommendation is going to be coming off the wall and not be counterbalanced by a lot of other people who have something different to say about it."[26]

Most presidents, of whom Reagan was typical, would not make a decision without first "staffing it out." There were, however, exceptions. President Kennedy, said one of his top aides, "was from time to time coming up with 'damn fool ideas' like going to the moon."[27]

Beggs had never been cleared to meet with the President on the space station issue specifically. He had brought the subject up, however, in at least one national security briefing on the Soviet space program and during Reagan's visit to Mission Control in Houston. NASA officials had at least one ally among the President's three closest advisers, Edwin Meese, and had planted the suggested remarks on the right occasion. If the President took the initiative and made a personal decision to proceed with the space station without extensive White House staff review, this would end nearly twenty-five years of waiting and would prove that only one vote counted in the Executive Office—that of the President. It would also, as one presidential aide observed, send the White House budget and science advisers into convulsions.[28]

Reagan walked though the construction hangar, underneath the wing of the spaceship Enterprise and out onto the speaker's platform. Mattingly and Hartsfield, who had changed from their spacesuits into NASA blues, stood beside him. The crowd cheered and waved their flags.[29]

"T.K. and Hank," the President said, "you've just given the American people a Fourth of July to remember."

Reagan had not yet begun the speech and already he had departed from the prepared text. "I think all of us, all of us who've just witnessed the magnificent sight of the Columbia touching down in the California desert, feel a real swelling of pride in our chests."[30]

"In the early days of our Republic, Americans watched Yankee Clippers glide across the many oceans of the world, manned by proud and energetic individuals breaking records for time and distance." The President read those words from his prepared text.

"The pioneer spirit still flourishes in America," he added. He likened the landing of the Columbia to the driving of the golden spike that completed the first transcontinental railroad. "It marks our entrance into a new era. The test flights are over. The groundwork has been laid," he said. "Beginning with the next flight, the Columbia and her sister ships will be *fully operational*."[31]

"Now we will move forward to capitalize on the tremendous potential offered by the ultimate frontier." If Reagan was going to endorse the space station, it would come at this point in the speech. With one sentence, he could simplify NASA's work and launch the initiative for which agency officials had waited nearly twenty-five years. "We must look aggressively to the future," he continued, "by demonstrating the potential of the shuttle and establishing a more permanent presence in space."[32]

A more permanent presence in space? Not only did Reagan ignore the space station, he avoided any commitment to a human presence in space. Hans Mark, when he delivered the suggested remarks to the White House staff, had insisted that the President at least include a reference to "the permanent presence of mankind in space." Without the human reference, the science adviser could interpret the President's remarks to mean that the United States could populate space with robots instead of people. A space station had to have people in it if it was to be a real settlement in space.[33]

"A more permanent presence in space"—the words were the product of the White House bureaucracy. The phrase was as much as Beggs could persuade the White House staff to accept, given the level of opposition to the idea and the natural reluctance of the aides to let the President commit himself prior to a thorough staff study. The language was meant to satisfy both sides, keeping open the possibility of a space station but committing the President to nothing at all.[34]

Reflecting on the speech after the event, NASA associate deputy administrator Philip Culbertson, who served as Beggs's primary adviser for the space station initiative, admitted that it was probably unreasonable for the agency to expect that Reagan would simply announce the space station program. "We had really presented him with relatively few facts," Culbertson explained. "I thought that it was quite unreasonable to believe that the President would endorse a program with that level of understanding."

"We knew that the program had to be thought through well enough that we could describe what it was, what it was capable of doing, what the probable schedule was, what it was likely to cost, and what its magnitude was before we could get him to support a program," Culbertson conceded. "In order for him to endorse a major new initiative by the agency, we had to have a reasonably solid story to talk to him about." That job would fall to NASA's Space Station Task Force.[35]

PART II

6
Budget Strategy

The speech at Edwards Air Force Base squashed whatever hope the bearers of the vision possessed that President Reagan would make an Apollo-type commitment to the space station. It provided fresh evidence that NASA officials would not be able to escape the slow grind of incremental politics in putting their new initiative on the national agenda. In the summer of 1982, NASA officials searched for the most advantageous way to carry their space station proposal past all the policy makers who would want to review it.

Even for a minor initiative, the review process could be exhausting, whereas the space station was anything but minor. "It is a quantum leap in space," said Edward Boland, chairman of the House Appropriations Subcommittee that had to approve NASA's budget. "We have had three of them: Apollo, the decision by Kennedy; the Space Shuttle, the decision made by President Nixon; and now the Space Station, which is another quantum leap in space, a quantum leap in effort, and also a quantum leap in dollars spent."[1]

In 1982, NASA consumed less than one penny out of every dollar the federal government spent (eight-tenths of one cent, to be exact). Outsiders viewed the space station as an effort by NASA to increase its "fair share" of federal spending and to do so at the expense of other agencies, many of which were being forced to take budget cuts in the battle to control the federal deficit. The space station proposal aroused the suspicions of the military. In 1982, for the first time since the civilian space program had begun, the military spent more money on space than NASA did.[2] Members of the defense community viewed NASA's space station as a competitor for the limited amount of money that was available to be spent in space. The space station proposal also triggered the concern of the intelligence community, which needed money for a variety of reconnaissance and communications satellites. It created anxiety among scientists seeking federal funds for research on non-space projects. It even agitated space scientists, who received NASA funds and feared that their own science projects might be sacrificed to help pay for another big space flight initiative. In an era of budget scarcity, fewer groups seemed willing to ignore NASA's request for one new initiative in exchange for noninterference on their own. One of the oldest rules of incremental politics—the assumption of reciprocity—was breaking down just as NASA officials were mastering the incremental game.

To compound their differences over spending, few of these groups could agree on the direction in which the U.S. space program should go. Some favored the high technology of automation, using robots and remote-controlled probes to conduct space research. Some wanted to promote the commercial potential of space; others wanted to emphasize earth resources, studying the planet from space. Still others wanted to send humans back to the moon and on to Mars.

NASA's space station proposal, offered during a period of budgetary constraint, invited the interest of a wide variety of participants, a broad "issue network" of agencies and interest groups that would be affected by any new space policy. They all wanted to participate in the decision, which was the primary reason why the White House staff had not encouraged President Reagan to make a quick Apollo-style commitment at Edwards Air Force Base.[3]

Outside interests preferred a comprehensive review of the space station proposal, a study group or special commission that could examine program alternatives and long-range goals. Given the level of disagreement over long-range goals, however, a study group could just as easily kill the space station as embrace it. Until the President established a long-term goal for the space program that dampened the warring factions, NASA officials were better off eschewing comprehensive review in favor of an incremental ploy.

Jim Beggs decided to advance the space station initiative as a small item in the agency's FY 1984 budget request. Known as "the wedge" or "the camel's nose," this budget strategy was one of the oldest maneuvers in the incremental game. An agency head would lay the foundation for a very large program by placing a very small wedge in the agency's base, hoping to minimize the scope of review by minimizing the size of the first-year appropriation request.[4]

Although it avoided the pitfalls of comprehensive review, the budget strategy was still a risky move. Opposition to the space station was growing, in large part because of the growing conflict over the budget. Every time the federal government took in $1.00 in revenues that year, it spent $1.35, adding to a deficit of mammoth proportions and exciting suspicion about any new demand on the expenditure side. NASA officials would have to put the space station initiative in front of budget officials who were telling other agency heads to cut their spending, neither of whom considered themselves true believers in NASA's overall vision of space exploration.

Beggs thought that the budget strategy could work. To prove it, he was prepared to appeal the wedge all the way to the President. NASA's budget request was due at the White House in mid-September. Through an elaborate process of staff review and reconsideration, White House budget officials would seek to forge a compromise with each agency head so as to produce a single set of numbers which the budget director could then take to the President. If budget officials and the agency head did not agree, the agency head could then appeal.

"It would be very rare for a director of an agency like the space program to go all the way to the President on a budget issue," said one official who had reviewed the NASA budget from the White House.[5]

"I thought that given an appeal all the way up to the President, which I was prepared to do, that I could get it approved at that time," Beggs said. "It was clear to me that the President favored the idea."[6]

Few officials outside NASA had done much thinking about a space station as of 1982. On the general political scene, the issue was nearly invisible. Except for the limited network of people who specialized in space policy, few knew that NASA even wanted one. To survive the budget review and position themselves for a presidential appeal, NASA executives had to tell people what a space station was. The image had to be specific enough to excite the interest of potential allies but broad enough to lead everyone to see some benefit in it for themselves. In the game of incremental politics, when people disagree on where they want to go, it is better not to show them the shape of the vehicle that will carry them there.

That posed a special problem for the people at NASA who were promoting the space station. In an agency full of model-builders and engineers, NASA employees liked to show off their designs. The architects of the space program had cleverly placed the main NASA headquarters building directly across the street from the site of the Smithsonian Institution's National Air and Space Museum. Museum curators displayed the results of eighty years of aeronautics and astronautical research, including the ships that carried NASA astronauts on their most famous voyages. The airplanes and spacecraft on display gave NASA an enormous advantage in a city where most federal spending seemed to disappear into a mysterious black hole. People could touch and see the results of space spending. Nowhere in the museum, however, could a visitor find a model of a space station like the one NASA officials wanted to build.

7

Wheels, Cans, and Modules

NASA officials did not want to debate space station design, because they had no design to debate. So many versions of the space station had been developed in the decades since the idea first appeared that no one could point authoritatively to one model and say, here it is. So fuzzy was the definition of *space station* that some people even suggested that NASA had already accomplished the mission.[1] If agency officials advanced any one design as the "official" NASA space station, they would lose the support of all the other people who held a different image in their minds.

Had a poll been taken in 1982, the average American would probably have said that a space station looked like a large, rotating wheel. That was the image Wernher von Braun had implanted with his 1952 *Collier's* article and it was the image Hollywood had reinforced in a number of movies.[2]

Von Braun's wheel measured 250 feet in diameter, turning on its axis every 22 seconds. A large, air-tight wheel of that size would have to be assembled in orbit, a demanding task, given the technology of the time. Von Braun's colleagues at NASA devised an ingenious solution to the orbital assembly problem, drawing up plans to launch the wheel with a single rocket blast. Since the largest rocket available on the drawing boards (the Nova, which was never built) measured 50 feet across at the base, a 250-foot-wide space station might hang over the sides of the rocket like an oversized bedspring on a pickup truck, creating some interesting aerodynamics during lift-off.

Von Braun and his colleagues addressed this problem in a unique way. They would inflate the space station. Using the principle of the automobile tire, they proposed to build the space station out of "flexible nylon-and-plastic fabric." NASA engineers at the Langley Research Center built an impressively large model of the inflatable space station, attaching the outer fabric to a central metal rim shaped like the hub of a wheel. Deflated, the hub and rim could be tucked into a single rocket of appropriate size. Once in orbit, the rubber bulb would be pressurized to create a space station that would rotate around the earth like a wheel on an automobile.[3]

Had this design gone forward, the Goodyear Tire and Rubber Company might have become one of America's largest aerospace contractors. It ran an advertisement for an inflatable space station in the *U.S. Naval Institute Proceedings,* boasting that "Goodyear Aircraft is in a unique position to put this new technology to work at once."[4]

Apparently the thought of a flat tire in space, with the commensurate problems of patching it while rolling along at 18,000 miles an hour, drove NASA to experiment with solid metal space stations that could also be tucked into smaller-diameter rockets. By far the most interesting concept to emerge from these studies was what NASA called the "radical module," better known as the Y-shaped space station.

Folded up, the Y-shaped space station fit inside a 33-foot-diameter rocket. Once in orbit, it unfolded like an umbrella being opened in the rain, its three large spokes, each 95 feet long, creating a space station 190 feet in diameter from tip to tip. The astronauts lived in the spokes. No wheel or outer rim united them, so to move from one spoke to another the crew had to pass through the central hub, where the artificial gravity created by the rotation of the umbrella disappeared.[5]

The movie industry improved enormously on these designs. One year before Americans landed on the moon, Stanley Kubrick released *2001: A Space Odyssey*. The modern phase of the movie opened to the sight of an earth-based shuttle gingerly docking with a very large space station, rotating in time to the sound of "The Blue Danube" as the earth passed by below. The rotation, slower than that of von Braun's wheel, produced a sense of gravity roughly equal to that felt on the moon. The space station measured 900 feet across and held an international crew of scientists, engineers, and bureaucrats from the various space agencies funding the facility.[6]

As the movie set box-office records, attention turned to Gerard O'Neill and his designs for the ultimate space wheel—very large space colonies. Constructed from materials to be mined from the moon and the asteroids, O'Neill's space wheels measured miles across, rotating in self-stabilizing points between the earth, the moon, and the sun. Designed to house ordinary people, not astronauts, space colonies featured trees and streams, parks and lakes, and charming suburban homes. One design, modeled after the city of San Francisco in its pre-megalopolis days, could gracefully house several hundred thousand people around the inside edge of a rotating cylinder 4 miles in diameter and 19 miles long. Large mirrors outside the rotating cylinder reflected sunlight into the interior through window panels that covered more than half of the surface area of the outer skin.[7]

NASA engineers, meanwhile, struggled to design a space station that could get off their drawing boards. All of the designs for wheels assumed that people in space needed artificial gravity, especially for the extended periods needed to establish orbital bases and conduct interplanetary missions. To satisfy the mathematics of centrifugal force, artificial-gravity space stations had to be large. Von Braun's wheel measured 250 feet from rim to rim, and it had to spin at the dizzying rate of three revolutions per minute just to produce one-third "g." Better gravity required bigger space stations, and bigger space stations cost more money.

Stations without artificial gravity, on the other hand, could be small. Outfitted as research laboratories in space, can-shaped cylinders no more than a dozen feet in diameter could be launched into orbit on top of medium-sized rockets like the Titan III. In fact, a small research laboratory could even be carved out of the rocket itself, with the fuel tanks removed from the upper stage to make room for a small crew and a few experiments. Floating through space, such stations would generate practically no gravity at all.[8]

When it became apparent that the race to the moon had ruled out a large orbital base, engineers at the Langley Research Center abandoned their plans for a large, rotating space station and began to design small cans. To make the cans small enough to meet budget constraints, engineers left off nonessential items like the airlock through which the astronauts would enter the orbital facility. In one version, the crew would dock with the research laboratory, seal their spacesuits, depressurize their space capsule, open their hatch, climb out into space, find the door on the space laboratory, open it, climb inside, pressurize the laboratory, and remove their spacesuits. The process had to be reversed for the return to earth.[9]

The Langley engineers commissioned studies on what they called a Manned Orbiting Research Laboratory (MORL). Designed to carry out a variety of biomedical, scientific, and engineering experiments, this "minimum size laboratory" appealed to research scientists, who saw the facility as a nice substitute for the more expensive rotating wheels. Some of the most interesting experiments to be conducted in space, such as the creation of germanium selenide crystals, required a gravity-free environment. Astronomers and other people who wanted to use the laboratory in space as an observation platform also insisted that NASA eliminate any unnecessary rotation.[10]

Some engineers clung wishfully to the artificial-gravity vision, reluctant to abandon the requirement that seemed necessary for long-duration missions such as a trip to Mars. One compromise, produced by the Douglas Aircraft Company, abolished the full-station rotation necessary for all-purpose artificial gravity but kept a one-person centrifuge inside the facility within which an astronaut could produce his or her own gravity for short periods of time by twirling about like a hamster on a wheel.[11]

The concept of a can-shaped station appealed to the U.S. Air Force. During the period when NASA concentrated on the voyage to the moon, the Air Force promoted plans for what it called a military manned orbiting laboratory, or MOL, which the press instinctively learned to pronounce as "mole," the small, burrowing insectivore with minute eyes and concealed ears which works in the dark.

The Air Force fully intended to use its 5,600-pound research laboratory as a "mole," a spy satellite from which a crew of two could monitor troop movements and verify conformance with nuclear arms agreements. It would, its advocates argued, give the Defense Department the "high ground" so necessary for military success. As advocates of the program within the Air Force proceeded to appoint their own astronaut corps, an opposing faction argued that the same functions could be carried out much more reliably with unmanned reconnaissance satellites. Facilities for soldiers in space, the critics argued, were expensive to maintain and vulnerable to attack. The Secretary of Defense eventually canceled the program, siding with the Air Force officers who preferred robots to people. The "mole" never flew, falling victim to the same sort of skepticism that would haunt NASA's space station proposal a decade later.[12]

NASA's ambitions for can-shaped space stations grew in proportion to the size of the rockets the agency developed. With the advent of the mammoth Saturn V booster, the rocket that shot humans to the moon, NASA engineers could propose cylindrical laboratories up to 33 feet in diameter. While NASA officials and aerospace contractors studied this design, the Space Task Group advising President Richard Nixon proposed that the United States link together a number of 33-foot-diameter space cans like the vanes on a pinwheel. When set in motion, this collection of space cans would create an artificial-gravity space base with room for fifty or more human beings.[13]

NASA's plans for a 33-foot-diameter space can never materialized, losing out to the higher-priority space transportation system. Instead of four space cans rotating around a common core, NASA had to settle for one 22-foot-diameter space can approved in 1967 as part of the Apollo Applications Program and rechristened Skylab. Large enough to serve as a laboratory in space but too modest to form the core unit for a permanent base, the facility could best be described as an experimental prototype for a future space station. NASA launched the facility on May 14, 1973, using a leftover Saturn V. From the outside, the cylindrical workshop looked like the third stage of the Saturn V, which in fact it was, outfitted for living and working in space. A telescope mount and four solar panels attached to the top gave the can-shaped cylinder the appearance of a windmill.

Over a period of six months, NASA flew a grand total of three missions in Skylab, logging just 171 days in space. After closing the door, the last crew boosted Skylab into a 283-mile-high orbit, where it sailed along in silence for six years, slowly loosing altitude until it fell in pieces onto Australia on July 11, 1979.[14]

NASA's Skylab did not cruise through space alone. In April 1971, the Soviet Union launched its own experimental space station, Salyut 1. If Skylab looked like a can, Salyut looked like a bottle. The crew ate and slept in the neck, commuting to work in the bottom. The Soviet bottle measured less than fourteen feet across at its widest point. Soviet space scientists replaced their Salyut space laboratory with a new model every few years, each similarly small but more sophisticated in design. Though never permanently occupied during its first fifteen years of flight, the Salyut facility allowed the Soviet space scientists to test the ability of their cosmonauts to endure long-term missions in space. In 1980 the fourth long-term crew set an endurance record of 185 weightless days, which approximated the time required for a fast trip from Earth to the vicinity of the planet Mars, one way.[15]

NASA gave its operational backup unit for Skylab to the Air and Space Museum, where it came to dominate the great East Hall. A tourist gazing up at Skylab from the floor of the museum might gasp at the size of the facility and be impressed by the potential dimensions of a number of 22-foot-diameter cans linked together to form a permanently occupied space base. Such a person could be forgiven for forgetting that the United States had stopped manufacturing rockets large enough to launch cylinders as large as Skylab as the race to the moon drew to a close. Any new space laboratories taken into orbit after Skylab had to fit into the 15-foot-wide cargo bay of the Space Shuttle. In the summer of 1970, one year after the landing on the moon, NASA engineers had accepted the inevitable and instructed the aerospace companies developing plans for a space station to forget the large-scale cans and confine their designs to modules less than 14 feet wide.[16]

This constraint brought NASA back full circle to von Braun's original concern. The crew for even a small space station, six to twelve astronauts, could not be packed into one 14-foot-wide module. To create a modular space station, NASA would have to launch a number of modules in a series of shuttle flights and link the modules together in space. The modules might not rotate, but they would have to be assembled in space. As President Nixon gave his approval to the space transportation system, aerospace contractors presented NASA with conceptual plans for a shuttle-launched modular space station.[17]

When the current generation of NASA engineers were growing up, A. G. Spalding & Brothers manufactured a popular children's toy consisting of assorted rods and wheels. Called Tinkertoys, the wooden rods could be inserted into modular wooden wheels to create fascinating structures, from windmills to ferris wheels, infinitely larger than the container in which the pieces arrived. NASA's modules and collapsible rods could be carried into space in the same way, in a cylindrical container 15 by 60 feet in size. In space, astronauts could erect the components into enormous structures 500 feet wide.[18]

Well before Jim Beggs kicked off the 1980s space station campaign, NASA officials had concluded that any future station would have to be assembled out of rods and modules delivered to orbit by the space shuttle. In spite of its large size, this modular space station would not produce artificial gravity. Entrepreneurs wanted a near-zero-gravity platform for materials processing, scientists wanted a steady platform for astronomical observations, and engineers responsible for stabilizing the lattice-like structure of modules and trusses wanted to avoid any unnecessary rotation. The Skylab experience seemed to suggest that astronauts could endure periods of weightlessness without permanent harm, especially if new crews replaced tired ones every 90 days.[19]

The resulting configurations of modules, trusses, and solar wings created a new form of space-based architecture. This was upsetting to people who had a different image in their minds. Why go to all the trouble of assembling a large structure in space that could not rotate? A writer for *Science* magazine complained that such a facility would not look like "the most famous space station in science fiction—the giant, rotating wheel in the film *2001*."

"NASA's station would not even resemble a wheel," the author continued. "Since it will have to be pieced together out of modules ferried up one at a time by the space shuttle, it will look more like something a child would build."

"Space City," the magazine cover read. "2001 it's not."[20]

In the twenty-four years since NASA's founding, its employees had studied literally hundreds of different space station designs. NASA engineers and their contractors had analyzed small stations, big stations, stations with gravity, stations without it, stations that looked like wheels, stations that looked like cans, space bases assembled out of cans, space stations assembled out of modules and trusses, and an endless series of variations between. In the first seven years after it was created, the Johnson Space Center conducted no less than thirty-seven major space station studies on different designs. Three field centers had competed to prepare the best plans, and two more had tried to get involved. In 1969, anticipating presidential approval of the initiative, both the Johnson and the Marshall Center had "gone to phase B" with their space station studies, the second stage in the definition and design work necessary to build a specific station. For nearly a quarter of a century, the space station had rested on NASA's list of long-term objectives. The agency conducted study after study, even though Congress never appropriated a dime to build a permanent facility.[21]

Since the late 1960s, however, one design consideration had remained constant. The space station, constructed out of modules and connecting devices, would grow incrementally. The size of the components might vary, as could the method of assembly, but the incremental approach did not. Any space station made out of modules could grow piece by piece. This was an advantage in a government where decisions also tended to be incremental, beginning from a small base and building up gradually into larger programs as new elements were added to an expanding core.

Incremental growth in space station design, however, created a problem. The variations permitted by piecemeal growth divided the NASA field centers. When Jim Beggs took over as NASA administrator, he found that the Marshall and Johnson field centers had gone down quite different paths in their decades-old efforts to design an approvable space station. Officials at both field centers wanted to build the space station, badly. They agreed that the space station would grow incrementally. Their search for the appropriately scaled space station had produced quite different designs. They disagreed fundamentally over how fast the incremental growth of the space station should take place and how big a base they should start with.

A certain amount of vagueness about space station design was an advantage so long as NASA engineers could give the public a general idea of what the space station would look like. The divisions within NASA, however, posed an entirely different problem. Different field centers were prepared to fight for different designs. If the Administrator of NASA went to the White House or Capitol Hill to present one type of modular space station, he ran the risk that representatives from the opposing faction within his own agency would undercut him by sending word to their friends that they preferred a different design and that the administrator did not speak for them.

Clearly, this was an intolerable situation. Given the strength of the opposition outside NASA, agency executives could not withstand carping from within as well. Until Beggs and the leaders of the Space Station Task Force got the people at their own field centers to sing from the same sheet of music, officials outside NASA would not take the space station proposal seriously.

8
Configurations

As of 1982, engineers at NASA's Marshall Space Flight Center favored what they called the "space platform" design, starting with open-faced pallets built to hold a few scientific experiments and adding on modules until people could live there as well. It was a modest proposal, cheap to start, and was calculated to appeal to the science community, whose instruments would fill the pallets. In the beginning, astronauts would work on the platform only when the space shuttle was docked to it. As more modules arrived, astronauts could stay longer, and eventually the facility would become a permanently occupied station.[1]

Wernher von Braun, who headed the Marshall Center from its founding in 1960 until he moved to Washington in 1970, never gave up on the space station idea. "Von Braun was always looking ahead," said William Marshall, who had been brought in by von Braun as an electrical engineer to work on the Saturn V rocket and had moved up to take charge of the center's space station studies.

Like von Braun, Marshall was a non-southerner at the Alabama center. He had grown up in Missouri, attended college in Oklahoma, and done graduate work in New Mexico and Indiana. In 1985 he sat in his office on the campus of the Marshall Center and explained how a center created for the purpose of building rockets had become involved in the space station effort and why it took the approach that it did.

"Dr. von Braun was always looking at how he was going to do planetary exploration," Marshall said. "To do planetary exploration you need low earth orbit where you can assemble and do your marshaling of materials and so forth and go on out. He always had the vision," Marshall recalled, "that he needed to get to low earth orbit." A space station in low earth orbit was von Braun's stepping stone to Mars.

The Marshall Center, like all of NASA's field installations, ran advanced development programs, think tanks where scientists and engineers sat around and thought up missions the agency might ask to fly in the future. "We had 33 foot diameter space station studies," Marshall recalled, looking back on the days when NASA sought missions that would build upon the race to the moon.

"We designed how we would build a space station utilizing 33-foot-diameter [modules]," Marshall said. "We would put it into orbit with the Saturn V vehicles. Then we would service it with the Saturn I-Bs and the Apollo spacecraft. That was our first space station attempt. Most of that was centered around a stepping stone which was to do manned planetary exploration—a trip to Mars, to be specific."

"When monies kind of dwindled down," the center pulled back to a 15-foot-diameter module that could fit in the cargo bay of the space shuttle. "We studied that, and as we studied it, the budget began to shrink and interest in space got smaller and smaller." Von Braun, by then, had left the center and gone to Washington, hoping to lobby for the mission to Mars.

"Finally, they said, 'Look, we're not going to do a planetary Mars trip, guys, you can forget about that.'" Officials in Washington told the engineers at the Marshall Center to stop development work on the space station and help build the space transportation system. All the engineers at the Marshall Center got in the way of a space station was Skylab, a prototype for a space station built out of a leftover third stage of the Saturn V rocket, one launch only. The Marshall Center built Skylab. It was responsible for practically everything but the astronauts.[2]

"Eventually the decision was that we didn't have enough money for a fully reusable shuttle," Marshall said. "We kept moving down the price," conducting phase A studies for a space station that would be small enough to fit into both the cargo bay of the shuttle and the agency's dwindling budget. When "that sort of fizzled out," Marshall continued, the engineers ratcheted their hopes for a space station down one last notch. "We decided that what we really needed to do, at least at the Marshall Space Flight Center, we needed to allow the shuttle to stay on orbit longer."[3]

"Our approach was to go with a free-flying module, which we later called the power module." Folded up, the power module would fit into the cargo bay of the shuttle. Once in orbit, it would spread its wings like a butterfly emerging from its chrysalis. Two large solar arrays, a source of electric power, would emerge from the sides. A radiator, needed to dissipate heat, would pop up from the top.[4]

Marshall described how the crew of the shuttle would use the power module to conduct scientific and commercial experiments. "Initially we started out by coming up with the shuttle, docking to it, plugging the power system into the shuttle, which would extend then the life of the shuttle in orbit," something the people at NASA's Johnson Space Center, which developed the orbiter, wanted to do.

Onto the side of the power module, using the orbiter's manipulator arm, the crew of the shuttle would attach an open pallet containing various scientific experiments. The orbiter would then return to earth, "leaving the payloads in orbit to operate 90, 120, or however many days that they were going to be up there to generate good science, and then come back later with the shuttle, change out the payloads, bring these home, put other pallets on, and so forth."[5]

"We were playing at our center all along with the scenario we thought we wanted to do for the agency because of the pinch," Marshall explained, referring to the budget constraints. NASA's plan all along was to get the power module into position, "which turned out to be about a $300–$500 million program, and then grow by leaving a manned habitat there, which would basically be manned only when the shuttle was there and then ultimately including enough life-support systems that you could leave man there and it would be a space station."

The manned platform, Marshall thought, could be built for "a billion to a billion and a half dollars, up to two billion dollars depending on how rich you wanted the habitat to be and how much science that you wanted to put on it." The platform would consist of a habitat module with room for a crew of four, a laboratory in which to do science, a logistics module, various open-faced pallets, and the power module.

Marshall, who by that time headed the center's advanced planning effort, sought support for his space platform from the science community. "I've learned that the way you sell programs is to find somebody who eventually agrees with you they need it and will help you, and not be pushing on the other end of the stick all of the time."

"I had plenty of science support," Marshall said. The national science committees and academics that advised NASA, he explained, saw his space platform as "a modest investment as far as the facility was concerned," leaving more money for science and the investigative side of the operation.

Even with this scaled-down approach, so drastically reduced from von Braun's original dream, Marshall could not win funding for the program. "The senators and so forth we briefed that came through here at that time got very familiar with what the power module was," he said. "We got that into a new start package in two consecutive years within the agency," he said, but NASA could not get it out of the agency into the President's budget proposal and onto Capitol Hill. Marshall blamed the "budget crunches" again and the fact that "the shuttle was trying to come on line at that same time."[6]

The NACA pioneers who put NASA together dreamed of ways to put humans permanently in space. They hoped that the early space flight program would leave the United States with spacecraft for exploring the heavens and facilities for conducting space operations. The old hands got their facility, devoted to the development of space flight. NASA officials staffed it with the best young engineers, scientists, and test pilots they could find. There was only one hitch. The Manned Spacecraft Center was not in space. It sat on the ground, on the salt grass pasturelands of Texas, 25 miles south of downtown Houston.

For nineteen years the engineers at the Houston center worked to get a piece of their facility into orbit. In 1963 they awarded the first major contract to the Lockheed Company to work on the "conceptual design of rotating space stations." In 1969 they argued, unsuccessfully, that the government ought to build a space station before it built a space shuttle. They kept pushing for a space station through the 1970s, spending money for studies but never finding enough support to get the program approved.[7]

From his office on the third floor of Building 1 at the Manned Spacecraft Center (renamed the Lyndon Baynes Johnson Space Center in 1973 to commemorate the politician on whose native soil it sat), Allen Louviere recalled his frustration in trying to get space users to support the space station idea. He had been working on space station studies for nearly twenty years, ever since he had first come to the newly formed center as a thirty-four-year-old engineer.

"There are some very powerful advisory groups around that advise this agency," he said, practicing his best Louisiana drawl. "They are like the leaves on the trees. They are all over."[8]

The "users"—scientists and entrepreneurs whose experiments flew on NASA spacecraft—gave the Johnson Center engineers lists of things waiting to be done in space. The engineers put the lists together in what they called a blue book. "It seemed like it was about two feet high," said Clarke Covington, who, like Louviere, had joined the center at its founding, when he was twenty-seven years old. Covington was a local, with two college degrees in mechanical engineering from the University of Texas.

"One of the things that made a real impression on me was that the science community at that time that gave us these big blue books, when we tried to get out of phase B into a new start, that constituency just disappeared. They didn't support the program at all," said Covington, "like zero."[9]

Covington and Louviere discussed the problem with Bob Piland and Max Faget, two of the top engineers at the Johnson Center. Piland wanted a space station very badly. Faget, said Louviere, "was very frustrated with the fact that every time we would try to do a big new advanced project we would get all balled up with the user requirements."

So the four agreed, "Let's don't build one that caters to the users; we'll build one that is an operational base, a facility, and then what we'll do is just let the users come on board and when they come on board they will have to make their own beds." So they designed a space station that, in Louviere's words, would be "a transportation depot in the sky." They called it a "transportation node."[10] Users would prepare their experiments and satellites for delivery to the transportation depot, where NASA astronauts would assemble and service them. Some of the satellites and platforms would be set out to fly in formation with the station, like jet planes around an airport. Many others would be assembled and sent out to geosynchronous orbit. "Assume with us," Covington and Piland wrote, "that the next ten to twenty years will bring requirements for large, complex space systems and that geosynchronous orbit clearly becomes a primary operational arena." NASA could use the space station as a transportation depot for sending people and machines to geosynchronous orbit and beyond.[11]

Covington explained how the spaceport would compensate for the limitations of the space shuttle. "We went through this rationale that looked at all the design limitations of the shuttle": it was limited to a "crew of seven, you could stay up a week or two, 65,000 pounds, low earth orbit, 5,000 pounds to geosynchronous orbit using the inertial upper stage at best and less when you used PAM-Ds."

"There was a lot of commercial work based around communications systems that wanted to be in geosynchronous orbit. We had looked at a lot of high-technology options that involved great big antennas that you couldn't carry in the shuttle payload bay, very expensive switching networks, things that you were probably going to want to maintain in geosynchronous orbit."

"We need a construction site in space, where we can aggregate building materials and put together something too big to launch on one shuttle launch." As part of their spaceport, the engineers designed "a little module with a manipulator or even an automated building device that would let us make big antennas that you could send out to geosynchronous orbit or any large solar array that you wanted to make for a big power system or anything like that."

To take these structures out to geosynchronous orbit, the engineers designed a reusable space tug. "We called that an orbit transfer vehicle." The space station, Covington said, "ought to support an orbit transfer vehicle." Payloads and propellants would be brought up from the earth to the space station. The crew would "prepare them for launch, check them out, make sure they were working. Then, with the space station just inexorably rolling around in orbit, when the launch window came around in low earth orbit you just launched out from there."

To provide "proper habitability," they put modules on the station for the crew to work and live in. They designed a logistics module that would be replaced every 90 days and would serve the crew "like a pantry."

They put a hangar on the station for docking and servicing satellites. Solar arrays jutted out from the station to provide power. Radiators dissipated heat. A docking mechanism connected the space shuttle to the station.[12]

Their space operations center, as they called it, was much more ambitious than the space platform being designed by the engineers at the Marshall Center. It cost four to five times as much, and that was just for construction, for the facility. That did not include the missions. Covington, Piland, and Louviere deliberately left off the major science missions. "In discussing all this," Covington announced, "we said, 'Well, what about a laboratory module on there for science?' "

"Remembering back at what had happened in the previous space station, we just sort of tongue in cheek said, 'Well, why don't we just not list that as a major objective of the space station now.' " Anticipating what would happen when they went out to explain their design, Covington predicted that "the science community will probably stand up and object and say, 'Those dirty rats left us off and why don't they put us on there.' And then we can say, 'You're right! If you are really all that interested we will make that a big objective of the space station and put it back on.' But it was really just sort of a tongue-in-cheek strategy to see if we could get that part of the requirements community to go on record as saying that they wanted to be included in the requirements of the space station, when before we'd been kind of left high and dry with that."[13]

Piland put it more directly. The scientists, he said, were not that enthusiastic about a space station. They would much rather have their own free-flying instruments, unmanned devices that the scientists could operate and control themselves. As for attaching scientific laboratories to the space station itself, Piland said, we "wanted what science was on there to want to be on there."[14]

9

The First Move

As soon as they took their story on the road, the leaders of NASA's Space Station Task Force learned how much of a handicap the internal disagreements over design created. Telling the space station story was part of the Task Force job. Along with Jim Beggs, Hans Mark, Philip Culbertson, and other NASA executives, Task Force leaders worked to spread the vision, describing the space station and their plans for it. People from Capitol Hill, the White House, other federal agencies, the aerospace industry, foreign countries, associations of scientists and engineers, and NASA advisory committees, as well as NASA employees working on other programs, heard the space station story. By themselves, John Hodge and Bob Freitag made more than eighty presentations in the eighteen-month period between the summer of 1982 and the end of 1983.[1]

In March 1982, before the Task Force was officially formed, Terence Finn had arranged a briefing for representatives from eight aerospace firms. The aerospace executives were sympathetic but skeptical. They were skeptical not of the need for the space station, Finn recalled, but of the way NASA was handling the proposal. "It was a criticism of us," Finn said. "NASA was criticized," the minutes of the meeting read, "for not being bold," for being "too defensive." Before NASA told its story, it needed one story to tell. "Different elements of the agency," the aerospace executives observed, were "pushing for different goals." They told Finn that "NASA needs to speak with one voice."[2]

The degree of division within NASA had surfaced the previous November in New Orleans, at a meeting called "to initiate the formulation of a coordinated agency program plan" for the space station. Ivan Bekey had called the meeting. As Director of Advanced Planning in the Office of Space Flight, Bekey was the person responsible for thinking up future missions for the space shuttle. He called the meeting on neutral ground, at NASA's Michoud Assembly Facility on the Mississippi River, before the Task Force was formed.[3]

Following Bekey's opening remarks, Clarke Covington presented the Johnson Space Center's plans for an orbital spaceport. Covington presented an ambitious, $8.5-billion space operations center designed primarily to service satellites, support spacecraft, and construct large systems to be pushed out to higher orbits.

William Marshall, representing the Marshall Space Flight Center, argued against the Johnson plan. "We at our center felt that we couldn't push that through Congress." If we tried, he said, "we'd just get laughs." Marshall recalled the unsuccessful efforts of his center to get any kind of space station approved, even the tiny power module. "If I could not get a $300-million program through, I did not see how I could get a $13-billion program through." The space operations center, he felt, would cost over $13 billion. "We had a better chance," he said, of "pushing through an evolutionary approach." In arguing for his space platform, Marshall noted that it was "something that had enough appeal to the science community, was small enough that we could put it under the tent and not raise the roof of the tent any." As for a fully developed space station, he said, the evolutionary approach "would ultimately get us there too and I thought would get us there sooner." Marshall left the meeting believing that his center was still on the right track.[4]

The twenty-seven persons who attended the New Orleans workshop did not resolve the design issue; the meeting was not set up to accomplish that. "A major objective of a space station is the support of science," the proceedings statement read, undermining the people from the Johnson Center, who left the science module off their space station plan. Moving on, the attendees agreed that the space station should support "a variety of space operations," admonishing the people at the Marshall Center, who wanted to think small.[5]

Robert Freitag, working at that time as Bekey's principal deputy, suggested that NASA develop a "mission model," a description of the activities to be carried out on the space station during its first ten years in space. "The justification for the space station cannot be solely based upon identification of potential mission users," the group responded in a half-baked way. On the other hand, they said, "neither is it felt that program approval can be obtained without preforming such analysis."[6]

The design problem, meanwhile, popped up in another forum. Jim Beggs, the NASA Administrator, formed a "blue ribbon" panel of outside experts to make an informal study of the space station initiative. He asked Dr. James Fletcher, who had headed the space program for six years under Presidents Nixon and Ford, to chair the group. Fletcher had considerable experience in securing new initiatives, having wrestled the space shuttle decision out of the White House in 1971.

Outside panels serve many purposes. Public officials use them to marshal support for a new initiative; panels can also provide useful advice. As Fletcher understood the charge, Beggs wanted him to find out whether the space station was "the right next step."

"Yes," the panel told Beggs, "it made sense, providing you didn't spend too much money on the first cut."

The second question proved more complex. Given that the space station made sense, "what kind of possible versions" should there be?

"We looked at a lot of that," Fletcher recalled.[7] The proposal for an unmanned space platform, the panel concluded, "did not make sense as a first step." People were moving into space, Fletcher argued—if not Americans, then Russians—and NASA had better set its sights on "a more permanent presence in space."[8]

Fletcher found the Marshall Center proposal for a space platform too modest; Beggs agreed. But if the Marshall plan seemed too modest, the Johnson plan for an operations center in space seemed "too grandiose." The Johnson people, Fletcher observed, "really had too many things they wanted to do with that space station."[9] Given "the current severe restrictions on the Federal budget," Fletcher noted, NASA ought to go for a "minimum manned platform" or "an intermediate manned platform which would have designed into it the ability to grow in size, power, and capability."[10]

Fletcher thought that NASA could win approval for a permanently manned space station that would cost about $3 billion, a price that was closer to the Marshall Center proposal than to the Johnson Center plan. The President's budget and science advisers, Fletcher believed, would sign on to a $3-billion space station and do so quickly. "We were worried about timing," Fletcher said, thinking about the budgetary constraints. "We thought that we ought to get going with something small soon."[11]

Fletcher's panel gave Beggs a number of ideas for a low-cost, quick-start space station. One of these challenged the notion that the space station core had to be delivered into orbit *within* the cargo bay of the space shuttle. A member of Fletcher's group resurrected a plan for carrying the space station aloft on the shuttle's belly. "No contractor or NASA Center," the panel wrote, "was looking seriously at the possibility of using expended external tanks as a possible platform." When the shuttle ripped into orbit, it took with it an external fuel tank that was 154 feet tall and 28 feet wide. "A number of inexpensive possibilities," the group suggested, "exist for using external tanks either singly or strapped together" as the basis for a space station. Not since the days of the Saturn V had NASA looked at a space station made out of building blocks of that size.[12]

John Hodge reviewed the report by Fletcher's panel, along with the other advice NASA was getting. Hodge agreed that the agency ought to put the first monies for the space station into the budget proposal that would go to the White House that fall. Hodge agreed with Fletcher that neither the Johnson nor the Marshall plan provided a suitable basis for starting a space station. At this point, Hodge believed that a space station proposal could be budgeted in the $3- to $5-billion range; such a plan would provide enough money to make the station permanently occupied, but it would not be as ambitious as the proposal from the Johnson Center.[13]

As for the specific design, Hodge took a new approach, one of the most important decisions the leaders of the Task Force would make. As a collection of engineers, NASA had a reputation for being preoccupied with design. Engineers liked to build things, so it seemed natural that when confronted with a possible new program, they rushed to their drawing boards or computer consoles to design it. "Hodge seems well aware of NASA's oft-criticized tendency to plunge ahead with hardware development before it thinks much about the users," one veteran agency watcher observed. "That impulse prevailed in the shuttle program," the writer said, "and as a result the agency spent the late 1970s in an embarrassing scramble for customers."[14] The decision to build the space shuttle had been plagued by an overemphasis on design, with the result that even economists in the Office of Management and Budget made design decisions. Hodge was determined not to get into a fight over design.

The intellectual catalyst for an alternative approach came from the agency's brain trust at Langley, Virginia. A short, jolly engineer from NASA's Langley Research Center by the name of Brian Pritchard joined the Task Force soon after its formation in May 1982. "I was at a point careerwise," Pritchard said, "where I thought I needed to spend a year in headquarters." Although he worked at the Langley Center, which had launched the agency's first space station studies more than twenty years earlier, Pritchard had never become involved in space station activities. "I had spent a number of years," he recalled, doing "mission analysis studies for planetary missions." Pritchard had also worked on the Space Shuttle Task Group. "I was on the mission panel," he said. "It was the first real definition of what the shuttle was going to do." Pritchard was "very intrigued" by the approach being taken by the Task Force.[15]

Donald P. Hearth directed the Langley Research Center from which Brian Pritchard took leave. A twenty-year veteran at NASA, Hearth had prepared a special management study for the incoming administration. Like many in the agency, Hearth was distressed by cost overruns on projects like the space shuttle and the damage they had inflicted on "NASA's credibility and reputation for successful project management."

Hearth and his fourteen-member study team interviewed NASA officials and examined a number of "representative" projects that had experienced changes in cost. His sample of cost growth ranged from a low of minus 4 percent (a savings over the original estimate) to a high of 127 percent (a major cost overrun). Costs started growing, Hearth concluded, right up front, when the project was first conceived. "Inadequate definition, prior to the NASA decision to proceed with project implementation," his group wrote, contributed significantly "to cost and schedule growth." In other words, when NASA rushed into a project without sufficiently defining what the project was supposed to do, it laid the groundwork for cost overruns and schedule delays down the road. Hearth recommended that NASA spend funds up front to carefully define the purpose of all new programs before asking for the money to build them.[16]

John Hodge took Hearth's recommendation and the person of Brian Pritchard and made his first major move. "To hell with the configurations," he said. "What are the requirements over the next 25 years?"[17] The Task Force, Hodge decided in the summer of 1982, would not attack the design question. It would concentrate on defining the missions the space station would perform and the requirements this would place on the facility. He took his approach to Philip Culbertson, who quickly approved it, throwing the weight of the Administrator's office behind the decision.[18]

"On the Task Force," Hodge said, "people got very mad at me, because of course there were a lot of engineers on the Task Force, mostly engineers, and they want to start designing things and drawing pictures." Hodge himself was an engineer.

"I wouldn't let anybody draw any pictures in that first eight months or so. Everybody was chomping on the bit," Hodge recalled. "Of course, there were always pictures appearing," he said. Working with their contractors, officials at the Marshall and Johnson centers had prepared slick art work promoting their proposals, and sketches periodically appeared in aerospace and science magazines. "But we never used pictures, we never showed that at all," Hodge said of the Space Station Task Force. "We concentrated for almost the first year on not what does a space station look like but why do you need it and what is it and what does it do."[19]

Beggs and Culbertson agreed that NASA ought to involve potential users in the identification of space station missions. Hodge gave this job to Brian Pritchard. Involve the members of the user community, he told Pritchard, by asking them to prepare mission analysis studies. Hodge found $6 million to pay for the studies, scrounging it from the office that oversaw the space shuttle program. The Task Force, an ad hoc organization with no official status within NASA, had no budget. Major General James Abrahamson, Associate Administrator for the Office of Space

Flight, gave Hodge the money. In turn, Hodge gave Abrahamson the credit for proposing the mission analysis studies.

"Abrahamson had said, 'You know, you need to mobilize industry.'" Hodge remembered Abrahamson's telling him to get some advice on the "fundamental basis" for the program, the reason NASA wanted it. The contractors would be asked to identify missions, develop alternative concepts, conduct cost-benefit studies, and list requirements, but they would not design any stations. Pritchard recruited eight aerospace companies and a dozen subcontractors to conduct the studies. NASA officials signed contracts for eight mission analysis studies on August 20, 1982, three months after the Space Station Task Force was formed. Old hands at NASA had told Hodge, "You'll never get those contracts out. It will take six months or a year to get those contracts out." Brian Pritchard pushed the NASA bureaucracy to get the contracts approved. "He had eight contracts on board in four months," Hodge observed, and "he didn't break a single rule."[20]

"The studies," the NASA announcement read, "are expected to identify and analyze the scientific, commercial, national security and space operational missions that could be most efficiently conducted by a space station. From this analysis, the contractors will develop alternative concepts for a station." Hodge took the design question out of the debate on the day the contract officers signed the papers approving the analysis of potential missions.[21]

It has been said that "the definition of the alternatives is the supreme instrument of power."[22] By deciding to focus on missions first and configuration details later, John Hodge defined the ground upon which the space station debate would take place. He removed the primary issue dividing the NASA field centers. He crimped the ability of outside groups to pick apart NASA on configuration details such as the question of whether the space station should be carried up in the cargo bay of the shuttle or be built out of the external fuel tank. Opponents lost the ability to shave down the size of the space station because Hodge did not give them an architectural plan from which to whittle. They lost much of their ability to criticize the cost of the space station because Hodge did not give them a picture they could price out. By framing the debate in terms of missions instead of design, Hodge ensured that the problems that had plagued space shuttle engineers a dozen years earlier would not afflict the space station planners.

At the time the shuttle was approved, budget analysts in the White House had whittled down the size of the system. As the price for approval, they had forced NASA executives to take a less sophisticated shuttle than NASA wanted and to accept cost objectives that were unrealistic. In the space station debate, Hodge effectively blocked that possibility, but he took a risk in doing so. His approach might save money, or at least hold out the promise of a better cost estimate. His approach made it easier for NASA to someday design a space station that would satisfy the needs of different users with, for example, different computer or thermal requirements. In the long run, however, his first move made it easier for opponents of the space station to undermine the program.

By insisting on an extensive period of mission analysis and requirements definition, Hodge guaranteed that NASA would not start to build the space station for at least three years. The analysis of missions and requirements would not end with the work of the Task Force. It would continue for at least two years. NASA would take two years to conduct more studies and issue a "baseline configuration"—a preliminary design. Members of Congress and various White House officials would want to study the studies. Then NASA and its contractors would spend an additional one to three years completing the detailed design work, or blueprints.[23]

Not until NASA officials actually "bent metal" would the program be secure. With all of the up-front planning, however, that might not happen for as much as five years. Enthusiasm for a new program tends to peak among politicians when they are called upon to approve it. Even if they initially approved the program, the President and Congress would get at least two chances to change their minds before actual construction began.

Lots of things could happen in that time. A new President could take office. The opposition party could win control of Congress. New crises could force revised priorities. At the least, enthusiasm for the space station would wane as other issues forced their way onto the national agenda. Politically speaking, it would be much better for NASA to get the money to build the space station at the same time that the agency asked for its approval. To do that, however, NASA executives would have to put the design question up for debate. So Hodge took it down, promising to spend up to 10 percent of the total program cost on up-front analysis and definition before the final blueprints were drawn.[24]

James Fletcher, who watched all this from a distance, said that Hodge's strategy was probably "a better way to go." If it worked, Fletcher observed, the nation would get "a real space station." He worried, however, that NASA would not receive permission "to go ahead and cut metal" in fiscal year 1987, in which case "it would have been better to go our way" and build what he called his committee's "quick and dirty" space station.[25]

In putting the analysis of missions before design, Hodge did not ignore the shape of the space station altogether. Working with NASA Administrator Jim Beggs, Philip Culbertson, and members of the Task Force, Hodge mapped out some rather specific ideas about the space station and how the program would be organized. "Every time we went to talk to him," Hodge said of Beggs, "he came up with another one." The leaders of the Task Force called them "boundary conditions," short phrases that explained the concept of the space station and gave contractors some guidelines on which to base their studies. Most of the guidelines were set down by mid-June, 1982. Hodge presented six "boundary conditions" on the shape of the space station, eighteen words in all.[26]

On one condition Beggs and the Task Force leaders would not compromise. The Space Station had to be permanently occupied. NASA officials had lived with that vision for nearly a quarter of a century and neither Beggs nor the leaders of the Task Force wanted to reanalyze it. They did not consider whether something other than a permanent human presence could serve their hopes for the future. "Permanent presence," the fourth guideline read, essentially rejecting the idea that NASA could do the same job with robots or human-tended facilities.[27]

The first guideline contained a terribly important assumption. "Shuttle compatible," it read.[28] Of all the different types of space stations that NASA had studied over the previous twenty-three years, only a fraction had to be carried into space in a 15-foot-wide container. Most of the early space station designs assumed that the space station would be rocketed into space on a larger launch vehicle. As originally conceived, the space shuttle was designed to ferry crew and equipment to the station, not to put the station in orbit.[29]

In order to prove that the space transportation system was cost effective, NASA had to increase the number of shuttle flights to the point that the spacecraft became the nation's primary launch vehicle for transporting all large objects into space. NASA officials could not propose the development of a new launch vehicle to take the space station into orbit. That would boost the cost of the station and call into question the purpose of the shuttle, which, at the time, provided the rationale for moving ahead with the orbital base.[30]

"Shuttle compatible" implied that the space station had to be built from a web of relatively small modules, which in turn forced NASA officials to envision a space station that would grow piece by piece. In Hodge's view, the space station would continue to grow for twenty-five to thirty years or more. As new modules and new technologies became available, they could be plugged into the old core. A modular space station would not resist growth as much as a big wheel would. The facility, Hodge wrote in two of his guidelines, would be "evolutionary in nature" and "user friendly."[31]

In order to satisfy NASA's broad constituency, the space station had to employ what Task Force leaders called the "potential amalgamation of manned and unmanned elements." Hodge listened carefully to Daniel

Herman, the fourth member of the team, who along with Hodge, Finn, and Freitag got the Task Force started. Herman's friends in the science community told him that people stomping around in a space station would interfere with many of the experiments the scientists wanted to carry out. A telescope with very precise pointing requirements, for example, could not have people bumping around it all the time. There were some experiments so delicate that the breath of the crew might foul them. Herman, who was the chief engineer on a team full of engineers, insisted that the space station not be a single thing. It should, he argued, consist of an amalgam of elements flying in a number of orbits around the earth. A central unit would house the crew, the service center, and a variety of onboard laboratory modules. It would be complemented by unmanned platforms flying from west to east in the vicinity of the primary core and from north to south in polar orbits. The vagueness over design, coupled with the promise of a space station program consisting of many parts, went a long way toward accommodating clients whose interests could not be reconciled on a single facility. When asked whether the space station would be manned or unmanned, the Task Force members could answer, "Both."[32]

"Autonomous operations," the last guideline read. Up until that point in time, every mission the United States had ever flown in space had been directed from the ground by crews working at NASA's Mission Control Center, three shifts per day, around the clock. John Hodge had served as one of the flight directors at the Mission Control Center during the Apollo era. He was determined that space station operations not be directed entirely from the ground. He wanted the crew to run the station, or at least part of it, from their bridge in space, both as a means of cutting the cost of ground-based operations and as an incentive for improving the sophistication of in-flight technology. This decision affected the design of the station. It meant that the crew would need its own mission control center on board. This imposed some tough requirements on the people designing the instruments that would monitor the performance of the station and the computers that would process and display that information.

The decision also had a social effect. It put the first nick in the umbilical cord that bound people in space to people on the earth. It rekindled the original vision of the station in space, which in Daniel Herman's words was to prepare "for the exploration of space" by men and women "in the 21st century." Perhaps people would move out to live and work in space, cutting their ties in some fashion to the humans who remained under the protective blanket of air on the earth.[33]

It was an inspiring vision, but for the present the leaders of the Task Force had their hands full of problems back on the ground. They had issued guidelines, but there was no guarantee that anyone would abide by them. The field centers still had their own agendas. The Task Force members had to figure out a way to make the guidelines stick. That required organization. The leaders of the team had to find some way to organize the Task Force so as to put an end to the internal differences within NASA.

10
How to Organize a Task Force

When they began their work, the members of the Space Station Task Force had little trouble making decisions. As the Task Force moved through its third month of operation in August 1982, it employed fewer than two dozen people, including clerks and secretaries, a small dot on the face of the federal administrative establishment. Members of a slender and cohesive team, Task Force employees could work together. If a member of the group did not cooperate, said Deputy Director Robert Freitag, he or she "got sent home in a hurry."[1]

Making the whole agency, with its 22,000 employees, speak with one voice was a different matter. That was a test of influence, a measure of the ability of Task Force leaders to win over people who worked in different parts of NASA and did not share the same sense of priorities. Official Washington tended to gauge the ability to exercise influence in rather simple ways. By those standards, the Task Force did not have much influence. It could not get telephones and it could not get office space.

In the third week of August, as official Washington dragged through its subtropical clime, members of the Task Force worked out of four offices and a broom closet on the fifth floor of the NASA headquarters building. Their quarters (suite 5131) were designed "to hold four professionals and three secretaries," the member of the Task Force overseeing administrative details explained. "We put in eighteen to nineteen people," Frank Hoban recalled. The sixth person to join the Task Force, Hoban was a twenty-year veteran of the space agency and an expert on purchasing, personnel reform, and productivity. He had served on von Braun's planning staff and George Low's cost-control project. Even though the Task Force reported directly to the Office of the NASA Administrator, "we couldn't get telephones." Hoban shoved four desks into one of the cubicles, informed its occupants that their quarters would also serve as a conference room, and told them to share one telephone. He could not get the agency to put four telephones in the room because the room was designed for only one person. Only John Hodge, Director of the Task Force, enjoyed the luxury of private quarters. The Deputy Director, Robert Freitag, tried to create an aura of privacy by installing a cloth screen between him and his secretary. One member of the Task Force, Lee Tilton, shared the broom closet with the Program Planning Schedule Board and the copying machine.[2]

The Spartan quarters brought the Task Force together, making it cohesive and giving it a degree of maneuverability that larger bodies lacked. These were advantages, but they were also signs that the Task Force had not yet reached the "critical mass" necessary to mold the NASA bureaucracy.[3] The Task Force was a small, off-line operation, not given much chance of success by the few NASA career employees who knew of its existence in 1982. The most talented people in the agency worked on other programs and did not want to have their priorities molded by a few dozen career employees working for an obscure task force. The leaders of the Task Force had to work with a fragmented agency; NASA had been designed that way. During the early years, the leaders of the infant space program had deployed their principal engineering and scientific talent to centers in the field, from Florida to California and north to Ohio. The leadership had in effect created nine "little NASAs" and placed them around the United States. The field centers, wrote one longtime NASA watcher, tended to plan "their institutional futures quite separately—more like 'friendly competitors' than full partners."[4]

Most of the agency's experts on space station planning sat at desks in Texas and Alabama. Employees at the Marshall Space Flight Center in Huntsville, Alabama, had received their final briefing on the space platform from the McDonnell Douglas Astronautics Company in May 1982. In January 1982, space station planners at the Johnson Space Center outside Houston, Texas, had been briefed by the Boeing Company and Rockwell International on the configuration for the Johnson space operations center. The centers relied on the contractors, the contractors competed with each other, and the competition rebounded through the agency.[5]

Officials at the Marshall Center planned to continue work on the power module that was the centerpiece of their space platform. Employees at the Johnson Center had invested three years of work on their space operations center. They took a very straightforward posture toward the Task Force. "They stonewalled us," said Bob Freitag. "They set up their own parallel organization, were off doing their own thing independently."[6]

The science centers were even less friendly. At the Jet Propulsion Laboratory, the Pasadena field center operated by the California Institute of Technology, employees devoted their lives to the exploration of space with robots and automated probes. They viewed the space station skeptically, as an orbital facility designed to eat up money that was needed to complete science projects, as the space shuttle had done. The reaction of the space scientists and engineers at the Jet Propulsion Laboratory, said their representative to the space station planning effort, was " 'We don't need it. It's a problem to us. It will just get in our way again. If we put our heads in the sand it will go away and leave us alone.' "[7]

In the long run, Captain Freitag hoped to get the best people from the field centers to come to Washington and serve on the Task Force. He wanted all of the centers to participate. The space station, he told them, "would be the mainline program for the next decade. It would be the centerpiece of all of NASA's development." All of the centers and all of the cultures within NASA "should benefit from it," Freitag said. "So they should have the opportunity of participating and contributing or extracting in a common way."[8]

The senior people at the centers, the ones who had been trying to get the space station started for ten or twenty years, did not want to go to Washington and work on the Task Force. Such a move would uproot their families and break up their careers. So Freitag, still needing their support, invited them to participate in a different way. He began to organize a series of working groups under the Task Force umbrella and gave most of the chairmanships to those people in the field.[9]

Officials at the Langley Research Center in Hampton, Virginia, had already sent Brian Pritchard to Washington. Langley Center engineers had initiated the first major space station study one year after the agency was formed. Freitag put their representative in charge of what he called the Mission Requirements Working Group, which would plan all of the missions to be conducted on the space station through the year 2000. Pritchard oversaw the mission analysis studies being performed by the eight aerospace contractors as well as the in-house working group that was organized to synthesize their results.[10]

To the Johnson Space Center, Freitag gave the chairmanship of the working group that would take Pritchard's missions and translate them into "requirements." The definition of requirements was the first step in the design of the space station. Pritchard's mission schedule, for example, showed that astronauts could capture and repair at least one co-orbiting satellite per month in the initial years of operation. This created the major "requirement" that the space station contain a sophisticated guidance and navigation system to assist in rendezvous and berthing as well as lesser requirements like the prohibition against outgassing of corrosive effluents "when the space station is in close proximity to the satellite." Pritchard's missions imposed hundreds of such requirements on the facility, from the demands for fire control to the problems of managing fluids.[11]

To the Marshall Space Flight Center, Freitag gave the chairmanship of the working group that would take those requirements and examine the different "configurations" the space station might take. The Task Force leaders insisted that the mission analysis and requirements studies take place first, before a baseline design was laid down, but the configuration group could consider the interplay between requirements and design, drawing up a few broad design alternatives without making a specific recommendation.[12]

To the Kennedy Space Center, which would have to launch the whole thing into orbit, Freitag gave the chairmanship of the working group that would think about the operation of the space station. To a certain extent, the operations group did the same work as the requirements group. It took the missions to be performed in the space station and translated them into requirements, albeit from the perspective of the people who would actually have to operate the facility. The space station program would create a mountain of operational challenges for NASA, from the first assembly of a large structure in orbit to the first transfer of large-volume propellants in space to lots of extra-vehicular activity.[13]

Freitag reserved the chairmanship of one working group for himself. Someone would have to figure out how to organize the program once it was approved. The aerospace industry would manufacture the parts that NASA eventually pieced together to build the space station, so a contracting strategy had to be worked out. NASA's field centers would supervise the contractors, but the job was so big that no one center could do it all. Potentially, Freitag felt, every center could be involved, so someone had to plan the division of work among the centers, a very sensitive task.[14]

The working groups drew their members from the field centers. A few people from headquarters were assigned to each group, but nearly all of the members came from the field. They served on the working groups part-time, from their centers, traveling to meetings and communicating via conference calls without moving to Washington, D.C. To coordinate their work, Freitag and John Hodge established a steering committee, made up of senior space station planners from the field. Like the members of the working groups, they served occasionally, did not join the Task Force, and thus did not have to move themselves or their families to Washington. Freitag described the Program Review Committee, as he called it, as a "kind of board of directors in which each center had a representative and each center sat in a council once a month and we reviewed the whole program. And we decided what the work should be for the next month and critiqued what was done the week and the month before. So the centers were deeply involved on that basis, right from the start."[15]

Freitag and Hodge knew that the working groups would have to do more than meet; they would have to produce something, a product that would make them disciples for the Task Force point of view. Each group, Freitag and Hodge decided, would produce a book, 40–400 pages long. Together, the books would describe the various elements of the space station: its missions, requirements, possible architecture, technology, and operation, as well as the overall management of the program. "It was [the] process," said Clay Hicks, the member of the Task Force who oversaw the production of the books. "Everybody participating, every center had an input," he observed. "It was the process that coalesced the NASA thinking that was all different."[16]

To assist the working groups, Freitag assigned members of the Task Force to help the groups prepare the books and finish their work on time. Typically, these were people who had careers at NASA headquarters: Claiborne Hicks, who came over from the shuttle office; James Romero, who had worked in the Office of Aeronautics and Space Technology; E. Lee Tilton, who came up to headquarters from the National Space Technology Laboratory; and Mark Nolan, who had served with Freitag in the Advanced Programs Division. Their assignments were fluid and changed from time to time. "John Hodge used a horizontal management technique," Freitag explained. "He didn't put everything in its place and organize with three or four people reporting; but he had fifteen, twenty-five people reporting to him. He would send a task off on procurement strategy, another task off on center organization, another task off on doing trade studies on engineering and so on." Hodge's horizontal organization "kept the base spread and where conflicts would arise between two of the groups working, well, he'd resolve the conflict." To maintain the fluid organizational policy, Hodge insisted that any member of the Task Force could attend any meeting or meet with the Director whenever he or she wanted.[17]

Freitag wanted the first editions of the books to be ready by December of that first year, 1982, less than four months away. Freitag explained the deadline and its relationship to the budget cycle. The first editions of the books would come out in December 1982 as the President made the final decisions on his budget proposal to Congress. The second editions, reaching more detailed conclusions, would come out in the spring of 1983 as Congress debated the President's proposal. Congress could appropriate funds to begin the definition and preliminary design work on the station that summer. The third and final drafts of the key books would be out as the fiscal year began in the fall of 1983, thereby providing the basis for the detailed planning that would follow.[18]

Freitag laid out two alternative budget milestones. The first showed "space station systems definition" beginning in the summer of 1983, the quick-start plan. The second put the definition studies off until February,

1984, a slightly slower start. Under the quick-start plan, Clay Hicks confirmed, the Task Force hoped to start spending \$3–\$5 million in phase B money in the summer of 1983, just one year away. Either way, the Task Force had to assemble its budget rationale in just one month, in time for the first round of budget review that would begin at the President's Office of Management and Budget in mid-September 1982.[19]

The Yellow Books, as members of the Task Force called them, would represent the all-NASA position on the space station. The Task Force gave them that name because that was what the Task Force that planned the space shuttle had put out a decade earlier: six books with yellow covers.[20] The Space Station Task Force, however, could not get the NASA bureaucracy to put out yellow books. Edition after edition, the books kept coming out with black lettering on white covers, gray covers, brown covers, and a color that would only kindly be described as orange. Even though the Task Force kept referring to them as Yellow Books in its presentations to outside groups, "they never were yellow until the final publication. No matter how hard I tried," John Hodge explained, "I just couldn't get yellow covers on them."[21]

That was not critical, however, since the Yellow Books were not designed for external consumption. When the leaders of the Task Force trooped out of their cramped quarters to make presentations on the space station, they did not take the Yellow Books with them; they made presentations based on the work that was going into the Yellow Books. The entire set of books, in final form, totaled 1,085 pages. Few things would have pleased the opponents of the space station more than to pin the Task Force down to 1,085 pages of written detail, which could then be analyzed line by line until the initiative died from protracted negotiation.

Earlier that summer, with the Task Force hardly one month old, the leaders of the team made another important move. They decided not to issue their decisions in written form, at least not in the type of writing that characteristically clogged the Washington bureaucracy. In one of his frequent strategy sessions with the leaders of the Task Force, Hodge asked if they could design a dozen transparencies for use on an overhead projector which would represent the group's thinking on the space station. The reply, Hodge recalled, was, "Would you believe twenty?"[22]

The viewgraphs became known among members of the Task Force as the "top twenty." In fact, Terry Finn, who handled external relations for the Task Force, prepared about one hundred viewgraphs. When someone had to make a presentation, Finn would assemble a set appropriate for the cause. Although some of the viewgraphs tended to appear in most presentations, such as the President's July 4th commitment to "a more permanent presence in space," others were picked out to appeal to the specific audience. Interchangeable viewgraphs required opponents of the space station to shoot at a moving target. When the position of the Task Force needed to be changed on a particular issue, the appropriate line in the old viewgraph could be excised and a new line inserted overnight. The Task Force hired a retired flight controller, Manfred "Dutch" von Ehrenfried, under contract to NASA, to manufacture and maintain the "top twenty."

As a group, engineers tended to express their thoughts in a form suitable for reproduction on transparencies. (The summaries from the eight contractor studies, for example, came back laced with phrases and diagrams in viewgraph form.) The Task Force, however, put its viewgraph production into the hands of a political scientist, Terence Finn, who raised the technique to an art form.

The most verbose of the viewgraphs barely exceeded 100 words, usually phrases in outline form. Whole sentences rarely appeared, except when Finn decided to quote someone like the President of the United States. Some of the viewgraphs contained almost no words at all. One, which showed the space shuttle Columbia landing, was labeled simply "now operational." Another presented a diagram of the Soviet Union's prototype space station with all the words in Russian. The strategy of brief phrases made the target small.[23]

Of all the viewgraphs Finn assembled for various presentations, the Task Force tended to use twenty of them over and over again. They proved the most persuasive and hence gave to the whole set the identifiable "top twenty" nomenclature. A "top twenty" presentation typically opened with President Reagan's new space policy. This was followed by some notes on the functions of the space station and the "reasons why." Finn then added a viewgraph, barely fifty words long and containing not a single complete sentence, which set out the current position of the Task Force on the shape of the space station program. "Shuttle compatible," this viewgraph read. "International participation. Evolutionary in nature. Agency-wide effort." There were some twelve guidelines in all.[24]

Following the guidelines, which were the centerpiece of Task Force policy, the "top twenty" proceeded through a series of charts that explained the organization of the Task Force, its working groups, the Yellow Books (titles only), and the schedule for completion of the Task Force work. A final set of viewgraphs typically described the five or six most important elements of the space station, such as the living quarters for the crew or the use of space robots or the potential for international participation on the facility. "As my friend Hodge would say," Finn remembered, " 'Next viewgraph please.' "[25]

The transparencies not only created a small, elusive target for opponents of the plan, but they were also important for what they did not contain. The "top twenty" made no mention of alternatives to a space station. They did not refer to the divisions within NASA over the size and shape of the facility. They contained no detailed specifications and they presented no approved design. In short, Finn did not allow a single exploitable topic to be represented in the "top twenty" viewgraphs. Had the Task Force sat down and written a typical government white paper describing its plans, it would have created a large, fixed target on which critics of the space station could have constructed a mound of controversy.

As the summer of 1982 progressed, the leaders of the Task Force continued to make presentations using the "top twenty," seeking to build a constituency. The meetings and presentations seemed to pay off. In late July, NASA's Advisory Council endorsed the space station proposal. Many government executives appointed advisory councils of distinguished professionals in their fields to consult with and advise agency officials. The membership of NASA's Advisory Council, however, included scientists, who could be expected to view the space station initiative skeptically. John Hodge briefed the Advisory Council as it met on July 21, 1982, using twenty-two viewgraphs, including five that had no words on them at all. NASA Administrator Jim Beggs told the Council that NASA would "propose significant additional funds" for the space station in the upcoming budget. "A permanent manned facility in space is not the only next logical step after acquisition of the Space Transportation System," the members of the Council wrote in their endorsement, "but is the most exciting prospect in space for the foreseeable future." They endorsed the effort to seek "a decision to initiate hardware development, hopefully in the near future."[26]

Freitag, meanwhile, continued to recruit more members from the field centers to serve on the Space Station Task Force. It was easy to get people from headquarters to join the Task Force, especially people who were coming off programs that were winding down and who guessed (correctly) that the Task Force might turn into a major new program with lots of job opportunities. Freitag, however, wanted the field centers to supply 70 percent of the people on the Task Force. "You're much more apt to get good people if you spread your base," he said. Moreover, he added, "you are building a constituency." As the working groups met and interest in the space station grew, Freitag found it easier to get the centers to send him good people, a sign of the increasing influence of the Space Station Task Force. Officials at the Johnson Center even sent up one of their top space station planners, Jerry Craig, to help run the Task Force. A constituency for the Task Force point of view was growing, not only outside NASA, but inside as well.[27]

II

International Participation

"I guess that's a memory jogger that will stay with me for a little while," said Dr. Karl Doetsch, the assistant director of the Canadian space program, remembering how the meeting to consider international participation on the U.S. space station ended.

"A meeting was called by Mr. Pedersen," Doetsch explained. As Director of the International Affairs Division at NASA headquarters, Kenneth S. Pedersen served as the principal contact between NASA and the foreign governments involved in the American space program. The meeting was scheduled at the Johnson Space Center in January 1982, a welcome respite from the Ottawa winter and one of the first steps in NASA's effort to build an external constituency for the space station. Doetsch arrived "to look at what was then called the space operations concept." At the briefing, Doetsch remembered, Pedersen "first addressed the question of potential international participation in the program."[1]

Doetsch was the project manager for Canadarm, the remote-control mechanical crane housed in the payload bay of the space shuttle. American astronauts used Canadarm to retrieve satellites and work in space. In exchange, Canada got to fly its payloads on the shuttle and would soon form a Canadian astronaut corps to conduct experiments in orbit. It was the kind of international participation NASA officials liked. Canada provided the equipment, and NASA launched it. NASA controlled the program, and Canadian engineers learned only as much about the technology of space flight as the United States allowed them to learn. The technical and managerial interfaces, NASA executives liked to say, were very clean.[2]

"We had a fair amount of information on the space station program," Doetsch said, through "the people that we worked with directly on the Canadarm." Doetsch knew that different versions of the space station were being promoted by people at the Johnson and Marshall centers. "We discussed what potential there was for the various international participants," Doetsch said, remembering that January 1982 meeting south of Houston. "I guess that was the first time that Mr. Pedersen indicated that they were seeking the international's input a lot earlier in the planning stages of that program than had been sought in the shuttle program. In other words, to be part and parcel of the concept development."[3]

The international participants, especially the Europeans, did not want to participate in the space station under the old terms. If they joined the venture, it would have to be as members of the crew with equal access to the technology that kept the station running. The attitude of the officials who ran the Johnson Space Center had been equally as adamant. It would be a cold day in Houston before they relinquished their technical or managerial control over an American space flight program. "That meeting," Doetsch smiled, recalling the event that continued to jog his memory, "was cut short as a matter of fact by a snowstorm." Freezing weather paralyzed the south. Sleet and snow fell so hard in Houston that NASA closed the Johnson Space Center.[4]

To gain an international constituency for the space station, NASA had to produce cooperation where there was no consensus. Since the program had not been approved, neither the White House nor the State Department would help NASA smooth out disputes. If NASA officials tried to resolve issues on an unapproved program over which the United States and its allies were divided, the agency would surely fail. Only by directing attention away from the divisive issues could NASA officials start to build support for an international space station.

In January 1982 the international participants were in a much better position to gain a larger role in the U.S. space program than they had been a few years earlier. Just one month before the Houston meeting, the nature of international cooperation and competition in space changed dramatically. On December 20, 1981, the European-built Ariane rocket completed its last successful test flight. Capable of lifting two conventional-sized satellites into geosynchronous orbit, the three-stage, liquid-fuel Ariane I gave the Europeans their own means for achieving routine access to space. The European nations that had contributed to the development of the Ariane I announced that they would no longer rely on the United States to launch their satellites and would in fact offer their own launch services for sale under favorable financial terms to any firms or nations that wished to purchase them.[5]

NASA's enabling legislation required its civil servants to cooperate "with other nations and groups of nations" seeking access to space, a seemingly altruistic goal. In approving the space transportation system, President Nixon had instructed NASA to fly foreign astronauts on the shuttle and to engage "in other types of meaningful participation, both in experiments and even in space hardware development." Armed with this mandate, NASA executives had sought international cooperation for the shuttle program. "Our fundamental objective," said NASA Administrator Thomas Paine, "was to stimulate Europeans to rethink" their objectives in space and "to help them avoid wasting resources on obsolescent developments." The Europeans learned a lot about NASA's definition of "obsolescent developments" through their work on the shuttle program.[6]

"It seems clear that our proposed space station/space shuttle systems would obsolete many of their proposed developments before they become fully operational," Paine argued back in 1969. By obsolescent developments, Paine meant the European plans to gain direct access to space with their own rockets. "The European nations have still not determined whether they should rely ultimately on cooperation with the United States or should develop a completely independent capability for space operations." NASA had placed all of its faith in the space shuttle as the technology of the future for routine access to space. It did not want to see the Europeans develop a competing launch capability based on expendable boosters. To interest the Europeans in using the space shuttle, NASA held out the possibility that foreign countries could participate in the construction of the system and build a laboratory that would fit into the shuttle's cargo bay. By scrapping their own plans and joining with the United States in developing the space transportation system, the Europeans could gain vital technical knowledge about space flight as well as extensive experience in space operations. They would be able to develop their own high-tech industrial base. They could operate as more equal partners in space. That, at least, was what the invitation to participate in America's space transportation system led them to believe.[7]

NASA officials held out the possibility that the Europeans might be allowed to build an orbital "space tug" as part of the space transportation system, a major technical challenge requiring extensive understanding of robotics and the handling of cryogenic fuels. The shuttle-fitted laboratory was less attractive, since it posed fewer technical challenges. To encourage European interest in the laboratory project, U.S. officials suggested that they might purchase several of the Spacelabs after the Europeans geared up their industries to produce them.[8]

While some groups within NASA wanted to hold out such promises, other groups worked to snuff them out. To bring the Europeans to the point where they could build a vital component of the space transportation system, the United States would have to share technical knowledge on which it held a monopoly. No other nation in the world had built a reusable space shuttle, and the attitude of many people within the U.S. government was that the country should not give this information away. Even if U.S. officials were willing to share technical information, they were universally unwilling to share U.S. military secrets. The space tug would have to handle military satellites, knowledge of which the U.S. Department of Defense feared could leak through Europe into enemy hands. From the U.S. perspective, only the Spacelab facility seemed sufficiently removed from the risks of unwanted technology transfer.

The Europeans did not have much room in which to bargain back in 1973. They agreed to build Spacelab with their own money and ship it to the United States. "Europe's most expensive gift to the people of the United States since the statue of Liberty," one German official remarked. The United States flew the Spacelab infrequently and purchased only one additional facility as required in the original agreement, disappointing many Europeans who saw international cooperation as a means to develop Western Europe's standing as a major astronautical power.[9]

Possibly with the money saved from not building the space tug, the Europeans proceeded with their own plans to develop the unmanned Ariane rocket. In the process, they also agreed in 1975 to form a single European Space Agency (ESA) with headquarters in Paris. As European plans for space exploration expanded, the eleven-nation consortium set more ambitious goals. Its members approved the development of Ariane II and Ariane III and listened to proposals for ever-larger versions that would give them the capability, like the Americans, to lift heavy objects into space.[10]

The French were just completing the flight qualification of the Ariane I rocket when Jim Beggs announced that NASA would make the space station its next initiative in space. Ken Pedersen approached Robert Freitag and Ivan Bekey, Freitag's boss at that time, to learn more about space station technology and to discuss the best method for presenting the international issue within NASA. Freitag was "the common denominator," said Karl Doetsch, Canada's representative. As special assistant to NASA's associate administrator for manned space flight from 1970 to 1973, Freitag had worked to generate the plan for international cooperation on the space transportation system. He and Pedersen discussed the advantages of international cooperation and the best way to enlist support from Europe, Japan, and Canada. They also had to worry about pockets of resistance within NASA to any kind of meaningful international cooperation.[11]

NASA executives needed allies to help get their space station initiative approved, a constituency of support that could counterbalance the opposition back home. If NASA officials could interest their foreign counterparts in the space station, that could be used as an argument to win presidential and congressional support. "Substantial foreign interest exists in NASA's space station," Freitag announced in one of his early briefings on Capitol Hill in the summer of 1982. "This interest derives from existing contributions to NASA's STS [and] past and present cooperative activities with NASA. Several countries have asked how they can be involved in NASA's planning."[12]

Implicitly—and sometimes explicitly—Task Force leaders pointed out the embarrassment the United States would suffer if it drew back from a program in which it had already generated substantial international interest. To withdraw would amount to an abrogation of U.S. leadership in space. To build international support, Task Force leaders encouraged international representatives to participate in the conceptual development of the program, much earlier than NASA had done in the past. At the same time, NASA officials kept reminding the foreign partners that neither the President nor Congress had approved the program. "We had absolutely no approved program whatsoever," Ken Pedersen reminded everyone. "We had nothing to invite them to."[13]

Freitag had urged Pedersen to make a presentation on international cooperation at the November 1981 space station workshop he and Ivan Bekey had scheduled in New Orleans. No international representatives would be present, just NASA employees and one representative from the Department of Defense, including a lot of people Pedersen thought would want to make the space station an "America first" program. "I had simply made the assumption that I would get some negative reaction or I would get some very hostile questions about technology giveaways," Pedersen said. "I got almost none of that."[14]

The participants at the New Orleans workshop had approved a two-page statement on international involvement. "NASA can derive significant benefits from international participation in its programs if they are properly structured and controlled," the statement read. The participants then proceeded to list the conditions under which those benefits could be derived. "Cooperation must be demonstrably beneficial to the U.S.," the guidelines read. "Foreign contributions should, insofar as possible, take the form of discrete hardware packages that lend themselves to clean managerial and technical interfaces." Each party would have to accept "full financial responsibility" for its share of the program. "There may be certain fundamental capabilities (i.e., systems, subsystems) that will be restricted to U.S. development." Cooperation in one part of the program "does not entitle each participant to technical information on the total project." Control of the overall design and development, finally, had to remain with the United States.[15]

The endorsement of international cooperation, Pedersen remembered, was viewed "almost as a throwaway." It was easy to see why. The guidelines gave nothing away. They reaffirmed the traditional, conservative values that had governed international participation within NASA for more than twenty years. "Put in somewhat exaggerated tennis terms," Pedersen observed, "NASA controlled access to the court and was used to playing singles."[16] Armed with the guidelines, Pedersen had proceeded to invite representatives from the European, Canadian, and Japanese space programs to the 1982 briefing at Houston. Though cut short by the snowstorm, the meeting lasted long enough for Pedersen to invite the international participants to get involved in the space station initiative and to gauge their reaction. The foreign participants, as Pedersen saw it, seemed reluctant and skeptical. They had followed the progress of the space station initiative closely and knew the opposition the proposal faced. The Japanese wanted to know under whose authority the meeting was being held. For whom was Pedersen speaking: himself, NASA, the President, or someone else? Pedersen explained that he was speaking for himself, though his sentiments on behalf of international participation were shared by NASA Administrator Jim Beggs.[17]

The international participants understood the resistance that Pedersen and Freitag would face just trying to convince their colleagues within NASA to engage in meaningful international cooperation, not to mention potential opposition from groups within the U.S. State Department, the Department of Defense, the Commerce Department, the Central Intelligence Agency, and the Arms Control and Disarmament Agency, each of which had an interest in protecting American technology and was even less willing than NASA to consider meaningful international cooperation on the space station.

The Europeans also understood what they could gain from meaningful participation in an American space station. In addition to the national prestige, the sense of exploration, and the experience of space flight, they would also gain access to the high technology fueling the American economy. That would be possible only if NASA agreed to abide by the principle of *mutual access*. Under this principle, each participant would have the right to use all of the facilities on the space station, including those developed by the other partners, subject to jointly agreed upon ground rules and a system of fees. Management of the space station, like the leadership of the European Space Agency, might even rotate.

Enthusiasm for NASA's space station initiative varied widely from country to country. The French seemed very reluctant. Given NASA's traditional philosophy of "America first," doubts arose about the willingness of the U.S. government to agree to meaningful cooperation. From the French perspective, the Europeans should go ahead and develop their own capabilities in space. They had less money to spend on space than the United States. Any money siphoned off onto an American space station would be like the money spent on Spacelab. It would leave the Europeans in a technological backwater.

Finally, there was the sticky issue of American defense. The international participants knew that NASA was trying to enlist the support of the U.S. Department of Defense for the space station. Military missions had been used as a basis for restricting European contributions to the shuttle program. The same thing could happen on the space station if the Defense Department came on board. European astronauts would need security clearance in order to walk through the military module, if they could get near the hatch at all. Furthermore, the international partners did not want to be part of a space station used to test weapons in space. It would destroy public support for their own civilian space programs at home. The Japanese were most insistent on this point. For political reasons, they could not participate in a space station program devoted to anything but the peaceful exploration of space.

Through the first half of 1982, as the Space Station Task Force started to take form, Freitag and Pedersen searched for ways to win support for NASA's space station among the Europeans, Japanese, and Canadians without getting mired down in these issues. Freitag thought that it was extremely important to involve the international participants, not just to build a constituency, although that was important enough. "In my twenty years of cooperation, I [have] come to feel that we all benefit by it," Freitag said. "We're not giving away the family jewels. And I think, personally—and this is not just for lip service—that it's important for us to learn to work together on a high-technology project of this scope, because someday it might be really important for us to know how to work together," he explained. "Whether it's a ballistic missile kind of thing, or saving the environment, or I don't know what, sometime we're going to have to work together on a real important thing like a large program," he predicted. "The language of the space age is English, and that's because they're following our lead. And, the time could have come when, had we not done this, the language of the space age could be Russian."[18]

NASA needed a method by which to involve the international participants in a way that got them excited about the space station without prematurely raising the issues that divided them from the United States. Freitag had suggested such a method at the New Orleans conference back in November 1981 when he urged NASA planners to focus on the missions the space station would perform. "Potential foreign participants," the group had agreed, "should be encouraged to fund and undertake parallel studies of Space Station requirements and concepts."[19]

In mid-1982, leaders of the Space Station Task Force involved American industries in their planning effort. The contracts for the domestic "mission analysis studies" were signed on August 20, 1982. Simultaneously, Task Force leaders encouraged their counterparts in Europe, Canada, and Japan to push for similar contractor studies at home. Ask your own domestic aerospace industries, proposed the Task Force leaders, what they would like to contribute to an international space station, assuming that one could be built under conditions acceptable to all participants. This strategy skirted the question of conditions, broadened the base of industrial support for the program, and involved the international participants in the same mission analysis process in which NASA was engaged.

NASA's counterparts, career civil servants working for other governments, did not have the authority to fund studies such as these. In Canada, Dr. Doetsch wrote a memorandum to the vice president of the National Research Council, indicating "that this is something we would like to explore further." His request for funds to finance the studies eventually reached the Canadian Cabinet.[20]

The Europeans responded first. At the conclusion of a meeting between NASA Administrator James Beggs and ESA Director General Erik Quistgaard in June 1982, the Europeans announced that they would study the possibility of participating in the U.S. space station program. Six months later they had their contracts out. Sixteen European industrial firms and subcontractors in West Germany, the Netherlands, Great Britain, Switzerland, France, Spain, Italy, Belgium, and Denmark received study contracts. The European Space Agency gave the primary contract "to assess the European utilisation aspects of a manned space station" to the German Aerospace Research Establishment—part of the West German government. Industrial contractors were instructed "to identify areas in which European participation in the US programme could be envisaged." Candidate programs included European participation in the generation of electric power for the station, self-propelled space tugs, space platforms, advanced pointing systems, and modules based on the Spacelab model which could be attached to the space station and manned by European astronauts. The study contracts encouraged foreign governments to think about the systems they could build rather than issues like access and technology transfer.[21]

The Japanese came on board second. Upon learning that ESA would participate in the study effort, Japan's Space Activities Commission requested funds for its own set of technical studies. At the start of 1983, the Canadian government announced that it had begun its own parallel studies as well. Since Canada had provided Canadarm for the U.S. space shuttle, it seemed natural to study the possibility of producing more advanced robotic systems that might roam around the space station and fix things in places where people would not have to go. At the same time, the government asked Spar Aerospace (the prime contractor for Canadarm) to look at other ways in which Canada could contribute to the space station. Look at the areas of remote sensing and communications, the government suggested, where the space station and its associated platforms could help Canadians monitor their huge land mass. Simultaneously, the National Research Council of Canada asked other Canadian industries how they might use the space station for communications, remote sensing, materials processing, and other space-based technologies.[22]

Given the level of suspicion that had built up between the United States and the international participants, the response of the Europeans, Japanese, and Canadians was surprising. Overseas industries (often consulting with their American counterparts) became as enthusiastic about getting their governments involved as did American aerospace firms. NASA put the question of participation to its foreign participants in such a way as to let their ambitions soar. It meant, as of late 1982, that U.S., European, Japanese, and Canadian space experts were ready to discuss plans for an international space station, even though none of the governments had approved the program.

Much had been accomplished. Within six months of its formation, the leaders of the Space Station Task Force had silenced the debate over configuration, enlisted the support of U.S. aerospace industries, organized the NASA field centers to participate in the planning effort, and convinced foreign allies to conduct parallel studies run by their own domestic firms. None of this had been done in such a way as to raise the issues that divided the parties to the decision. Those issues would have to be addressed someday, but not immediately. The leaders of the Space Station Task Force were preoccupied with getting the program approved.

Obscuring differences by allowing parties to believe that a single program will serve many purposes is a common incremental technique. In pursuing a comprehensive decision, such as the decision to go to the moon, parties to the deliberations must be specific about their goals. This brings differences out into the open. In the Apollo decision, of course, President Kennedy was willing to use his influence to resolve those differences by defining the ultimate goal. No such requirement, however, need motivate an incremental decision.

Pushing from below against a mantle of political disagreement, NASA officials had no choice but to put off the really tough issues. In one area, however, NASA officials refused to restrain their vision as a price for getting the program approved. In discussing the technology that would drive the space station, NASA scientists and engineers abandoned incrementalism. In planning space station technology, NASA officials let their real instincts come through.

12
Technology

"At some point in time we are going to do a manned Mars mission," said Danny Herman, the Task Force leader who would eventually become director of engineering for the space station program. "The space station is going to be in effect a test bed to evaluate what are the major issues before we can seriously consider manned planetary exploration."[1]

Richard Carlisle would eventually become Herman's deputy in the space station engineering division. In 1982, however, he worked for the NASA office that commissioned the technology research programs of the future. A well-dressed gentleman with degrees from Columbia University and Yale, the soft-spoken Carlisle looked more like a corporate chief executive than a space flight engineer.

"When we started pitching the budget augmentation," Carlisle explained, referring to the funds NASA would request for the study of space station technology, "we didn't have a friendly audience necessarily and the people who were opposed to the space station came at us and said: 'Hey, the Skylab was a space station and people stayed up there two to three months. We could have kept the Skylab program going if we wanted to; we knew how to do that twenty years ago. What do you guys need advanced technology for?' " Many opponents of the program, inclined to spend as little money as possible on the facility, pictured the space station as nothing more than a small research laboratory like Skylab. "The Russians were in orbit already with a space station," Carlisle observed—a small, single-module orbiting laboratory. "You mean to say," he heard the critics comment, "that U.S. technology doesn't compete with the Russians."[2]

In trying to describe the technology necessary to build the space station, Herman and Carlisle encountered the same ambiguity that had plagued NASA planners for twenty years. No strict set of criteria defined what a space station had to be. It could be a modest research laboratory in space or it could be a high-tech transportation terminal that pointed to the stars. NASA's engineers had drawn up plans for both types and in 1973 had flown a laboratory-type facility called Skylab, which they abandoned the following year.

In 1971, the Soviet Union had launched the first in a series of small orbiting laboratories called Salyut. Soviet engineers took a conservative approach to space station design and evolution. Cautious about the technical difficulties of constructing permanent facilities in space, they spent fifteen years working on their Salyut series, experimenting with girder assembly, robot resupply ships, power augmentation, refueling in orbit, and crew endurance. Not until 1986 did they launch the first module containing a multiple docking port (named Mir), the most elementary piece of equipment necessary to link modules together into a real working space station. Having launched the multiple-docking-port module, the Soviets then brought the crew back to earth before attempting to link the first laboratory module onto it. At no time during those first fifteen years did the Soviets attempt to make their space station permanently manned.[3]

Herman and Carlisle did not want to spend fifteen years plodding along with a small can in the sky. As NASA had done in the Apollo and Shuttle programs, they wanted to take a quantum leap. They wanted to begin with a multi-module, permanently manned, high-tech space station without any intermediate experimentation.

Hans Mark, Deputy Administrator of NASA and the person for whom Herman worked during the pre–Task Force period, told a group of NASA engineers and scientists in December 1981 to take risks when it came to technology. "The space station is inevitable," he predicted, "and we should be prepared in the likely event this administration, towards election time, may do a turnaround and want 'something big.' "[4]

"You can build a space station that has no engineering risks at all," Herman admitted, but it would not be the space station of which he and Mark and Carlisle dreamed. "In lots of subsystem areas we are going to have to make [the] decision as, 'Do you design for enhanced capability to some degree of risk or do you take the safe approach and stick with today's technology that you know will work, and there will generally be no risk?' "[5]

"Here I'm speaking for myself and not for my colleagues," Herman explained. "The space station per se as a low earth orbital vehicle, while it will provide the capability to have a permanent presence in space, that engineering endeavor, doing that, is not exciting to me. What excites me about the space station is it provides the capability to serve as a staging base for the start of manned exploration of the planets. Mars or a base on the moon. You couldn't undertake any of those endeavors without a space station. And my excitement in the space station is the potential to do just that."[6]

Mark, Herman, and Carlisle sought a method by which to spread the commitment to a quantum leap space station throughout NASA, where engineers and scientists at different centers still held quite different views about how big a space station the agency should try to build. To achieve their vision, they asked a question that NASA's engineers and scientists could hardly refuse to answer.

Well before the Space Station Task Force was formed, back in 1981 when Daniel Herman and Terence Finn were the only people in the Administrator's office working full time on the space station proposal, Jim Beggs and Hans Mark called in the director of NASA's Office of Aeronautics and Space Technology and asked him to answer a seemingly benign question. That person, Walter Olstad, worked for Beggs and directed the office at NASA headquarters that was charged with developing the technologies necessary to conduct new missions in space. An engineer with a Ph.D. in applied mathematics from Harvard University, Olstad had a $368-million budget to spend on space technologies of the future.

Beggs and Mark asked Olstad to put together a small steering committee composed of experts from NASA's field centers, people who held diverse viewpoints on the space station proposal, and challenged them to answer the following question: "[Are there any] high-payoff technology options which must be pursued aggressively to provide a solid foundation for a future space station initiative?"[7]

Asking a group of NASA engineers whether the agency needed to spend money developing new space technology options was like asking a convention of dentists whether people should get their teeth cleaned more often. Good engineers like to engineer. When assembled in conference, they exhibit an unfailing tendency to support the most advanced technology and the funds to develop it, whatever differences might otherwise divide them.

Olstad convened a twelve-person steering committee which met for the first time on December 10, 1981, five months before Beggs officially established the Space Station Task Force. From the Marshall Space Flight Center, Olstad summoned William Marshall, who had been working on space station studies for fifteen years. From the Johnson Center in Texas, he invited Allen Louviere, who had twenty years of experience in space station analysis. From the Langley Center, he brought in Paul Holloway, a twenty-one-year veteran who ran the research center's space programs. In all, the steering committee contained representatives from eight field centers plus three people from Washington, D.C. To organize the committee and serve as its executive secretary, Olstad chose Richard Carlisle, who at that time managed Olstad's spacecraft technology programs.[8]

Olstad asked each committee member to pick the "top 10 technology challenges" that would confront the agency during the initial design of the space station and "the longer term technology to be used for later application for improved capabilities." Carlisle tabulated the responses and presented them for discussion at the January 1982 meeting.[9] The space station, the group agreed, would need a steady, reliable source of electric power and lots of it. The space station would need a life-support system that could at least partly renew used water and air in space, both for the crew inside the station and for the astronauts working in spacesuits around it. And the space station would need a highly sophisticated data management system, computers to monitor the automatic functions of the station and help the crew perform daily operations.[10]

Unlike previous spacecraft, the space station would generate "huge quantities" of heat, primarily as a result of the generation, transfer, and storage of electric power. Radiators would have to be designed to dissipate that "waste heat" into space before the astronauts could throw the switch that turned the space station on.[11]

Once assembled, the space station would start to fall back to earth, for "the size of a manned space station and its necessary power system will produce large drag forces" in an orbit no more than 300 miles above the surface of the globe. To counteract the effects of drag, the space station would need engines and propellants, simple non-corrosive propellants that could be transported up in the space shuttle and off-loaded onto the space station, a tricky business given the technology of the day.

Before it fell out of orbit, the space station might shake itself into pieces. The engineers on the steering committee approached this problem gingerly. "Any inherent nonrigid characteristic of the space station," one member wrote in the jargon of the day, "can result in multiple natural modes of flexibility with frequencies within the attitude and autopilot controllers' bandwidth." "A dynamic interaction between the structural behavior and the closed-loop control response," he struggled to explain, "can thus result which can lead to basic instability and loss of the control function." Unlike space stations of the imagination, which tended to be rigid wheels and cans, NASA's modular space station would hang like clothes on a laundry line, creating "attitude, control and stabilization" problems of a magnitude the agency had never faced before.[12]

Like a building growing old, the space station would inevitably sag and bend if not constructed out of long-lasting materials. The lengthy girders on which the station modules would hang could deteriorate, causing misalignment of various parts of the facility. Structures and mechanisms had to be developed that would resist long-term decay.

NASA's communications technology did not match the needs of the space station. Communications equipment would link the station with the shuttle orbiter, space tugs, space cranes, data-relay satellites, global-positioning satellites, free-flying satellites, co-orbiting platforms, and astronauts outside the facility on EVAs. "The number and the wide variety of links required," wrote committee member Paul Holloway, "suggest a telecommunications system complexity unmatched in previous spacecraft."[13]

The group also felt that NASA needed to study "systems and operations" technology, the rather basic problem of making sure that any one piece of the space station could be plugged into all of the others. This problem was made especially difficult by the fact that the space station, designed to operate in microgravity, could not be fully assembled and tested-out on the surface of the earth.

To round-off the top ten technologies, the steering committee added a line on "space human factors." This seemed like a nice gesture toward the astronauts, who had complained in the past that engineers tended to squeeze them into their spacecraft as an afterthought. In fact, this was an afterthought. The committee initially wanted to focus on navigation technology, and only after five months of reconsideration confessed to the need to study the role of people on a permanently occupied space station.[14]

With these "top ten technologies" in hand, Olstad and Carlisle proceeded to expand participation on the committee. They established ten discipline-oriented working groups, one for each technology challenge, inviting new members to attend a meeting of the steering committee scheduled for March 15–18, 1982, at the Marshall Space Flight Center in Huntsville, Alabama. More than one hundred NASA engineers and scientists answered their call. Olstad and Carlisle asked them to prepare a technology plan for the space station and have it ready by June 1982, three months hence. The working groups returned with more than one hundred suggestions for technology research in support of the space station, projects that would complement or redirect research already under way.[15]

"Probably the biggest technological challenge that we're pursuing is in the power area," Danny Herman said. The Russian Salyut space station had limped along with 2–4 kilowatts of electric power, only enough to operate an ample American home. NASA's Skylab had been designed to generate 24 kilowatts of power from six large solar blades. One of the arrays, however, had ripped off during the launch and another had to be unfurled by the astronauts struggling in their spacesuits with inadequate tools. Given all the problems and inefficiencies in the system, the solar arrays on Skylab generated about 7.5 kilowatts of continuously usable electric power.[16] "We would very much like to have the initial space station have a solar dynamic system as opposed to a photovoltaic system," Herman dreamed. Both the American and the Soviet orbiting laboratories used photovoltaic systems, solar panels made up of silicon cells that converted sunlight directly into electricity. A solar dynamic system, on the other hand, could generate electricity the way it is done on earth, using heat to drive a turbine. The heat would be collected from the sun, using reflectors that worked like a lens. As the space station cruised across the sunlit phase of the earth, heat from the sun would melt a special material behind the collectors. On the dark side of the earth, the material would change back to a solid state. The melting and solidifying of the material would provide a continuous source of energy to drive the turbines.

"A photovoltaic system would be very difficult," Herman explained. To generate the electric power necessary to operate a truly ambitious space station, the solar arrays would need to be huge. Even at orbital altitudes, Herman made plain, "you would get aerodynamic drag." With a solar dynamic system "you would need a lot less square acres of frontal area for large amounts of power, and you could easily go up to two to three hundred kilowatts," enough to operate a truly sophisticated space station. "The problem is the engine," Herman explained. Either the engine had to be able to operate at high temperatures, or NASA had to develop a material to fuel the engine that melted in sunlight and solidified as the station swept across the dark side of the earth. "With a material that's non-corrosive, non-toxic," Herman promised, "the system will last."

Herman wanted a space station that would last. "The space station is not just going to be up for a week like the shuttle and come down to be repaired," he said. "It is going to be, in effect, like putting a Mount Palomar Observatory into earth orbit. It's going to really be a permanent facility." This created many new technical challenges. "We've never been posed with the problem of putting something as large and complex up there as a permanent facility."[17]

To make the space station more nearly permanent, NASA had to learn how to cut the umbilical cord that bound the facility to the earth. At the most elementary level, that required at least partial closure of what NASA engineers called the environmental control and life-support system (ECLSS). On the Skylab mission, NASA had relied on the brute-force, pack-it-all-in method of life support. Twenty-two steel and titanium tanks mounted around the airlock and the upper rim of the forward compartment had carried all of the oxygen, nitrogen, and water that three crews of three astronauts needed during their 171 days in space. The Soviets had resupplied their Salyut stations with automated Progress resupply vehicles launched into space on top of unmanned rockets.[18]

For short-term missions in low earth orbit, air and water could be hauled aloft in logistics modules, either in the shuttle cargo bay or on top of rockets. For really large space stations, and future bases on the moon or trips to the planets, NASA engineers had to learn how to create self-sustaining little worlds.[19]

At the time that Jim Beggs committed the agency to the pursuit of the space station initiative, NASA was already working on what its engineers and scientists called "closed loop" life-support systems. Water vapor could be recovered from the cabin atmosphere by simple condensation techniques. Wash water could be recycled through ultrafiltration. The astronauts could extract water from their own urine if they distilled it, and using an advanced oxidization technology, they could even gather water from what NASA engineers bashfully termed "organic solid wastes."[20]

If water could be recycled, so could oxygen. "Cells now under development," NASA engineers observed, "can supply oxygen with good efficiency by extracting water for electrolyzing directly from the cabin atmosphere." A two-step process could even manufacture oxygen from the breath of the astronauts. A special machine could gather carbon dioxide from the cabin, change it into water using a chemical catalyst, then produce oxygen by passing an electrical current through the liquid.[21]

"A fully closed ECLSS will not be feasible in the time frame" for the initial space station, wrote Paul Holloway, the Langley Center engineer and physicist who had been working on space technology for more than twenty years. "Partial closure of some loops," he predicted, "is possible."[22]

Engineers and scientists studying space station technology identified research challenges large and small. NASA's spacesuit, for example, needed to be redesigned. Existing spacesuits were bulky and awkward. They required the astronauts to sit for three hours and breathe oxygen in order to purge nitrogen from their blood and other bodily fluids. Worse still, the existing spacesuits used a cooling system that vented about two pounds of water for each hour each astronaut worked in space, creating little white clouds of vapor that could follow the space station in its orbit about the earth.[23]

Engineers working on the heat problem wanted to accelerate research on what they called a "two-phase thermal buss." Allen Louviere explained how the space station would dissipate heat. "Most radiators," he said, "have a fluid system like in your car." Known as single-phase thermal systems, they use power to pump a liquid fluid (like antifreeze and water) through a large, bulky radiator that gets rid of the heat. "Single-phase fluid systems," Louviere argued, "are very heavy and you have to pump a lot around and it takes a lot of power." Two-phase thermal systems, on the other hand, weigh less, take up less space, and consume less electric power. NASA engineers knew how to build single-phase radiators; they were still experimenting with two-phase heat pipes and capillary systems.[24]

Quite a number of the members of the working groups wanted NASA to pursue what they called "an integrated hydrogen-oxygen system." The space shuttle could arrive in orbit with its large external fuel tank still attached. The tank was not empty; it still contained some fuel. Normally, the shuttle crew jettisoned the external tank and let it fall back to earth. If that excess hydrogen and oxygen could be kept in orbit and loaded onto the space station, it could provide the crew with propellants, air for breathing, a secondary source of electricity, and fuel for space tugs and deep space probes.

"Adoption of an integrated hydrogen-oxygen system," the steering committee wrote, "offers the potential for minimizing or even eliminating many other space station consumables." One member said that the technology involved "promises the greatest reduction in resupply costs of any single development." Promising in concept, the technology posed a major engineering challenge. Cold, explosive, highly pressurized liquid hydrogen and oxygen were difficult enough to move around on the ground; in space they would prove much harder to transport.[25]

The steering committee identified the integrated hydrogen-oxygen system as one of seven major "technology themes," what committee members called "common threads of technology" that affected the work of all scientists and engineers studying space station technology. The "technology themes" endorsed the desire of Herman, Beggs, Mark, and Carlisle for a highly sophisticated space station. They endorsed the need for an advanced information system to help run the station, an energy-management system that could produce increasingly larger amounts of electricity, an operational philosophy that would "reduce the number and cost of ground support personnel," and automation techniques that would make the space station ever more autonomous.[26]

"When we first went into space, the fundamental design approach was to keep all the complication on the ground and put the simplest vehicle in orbit," Carlisle explained.[27] The old technology allowed flight controllers, sitting at their consoles in Houston, Texas, to operate a space station from the ground. Controllers on the ground flew Skylab for ninety-nine days without a single crew member on board. New computer technology, wrote chairman Olstad, "can eliminate most of the need for flight control operations now carried out at mission operations centers." Other members agreed. "Achieving an increased level of on-board autonomy and automation" was placed at the top of committee member Paul Holloway's description of technology themes. Herman and his fellow members on the Task Force cheered this call for "autonomous operations," as they called it. "Decoupling most of the crew activity from ground control," they announced, was one of their major technology goals. The key to accomplishing this, the technicians wrote, rested "in the ability to initially deploy the space station with massive excess capacity," computers with extra capacity to store and process data, computers with backup lines for transmitting signals, computers that might even be programmed to learn.[28]

Computers on the space station would control attitude, stabilization, soft docking, refueling, rendezvous, satellite deployment, and satellite retrieval. They would assist in laboratory experiments and oversee space manufacturing. They would be used for extra-vehicular activity (walks in space), space construction, and the generation, storage, and distribution of electric power. They would run the life-support systems and operate the automated platforms flying in formation with the space station. To make the space station autonomous, those computers had to be reprogrammable and self-correcting. Not only would it be necessary to give the crew the capability to reprogram the computers in space, but the computers would have to contain the ability to recognize and correct their own errors or survive a power failure with their memory intact, "preferably without intervention by on board crew members."[29]

Herman was familiar with the arguments about how this would save money by reducing the cost of ground operations, but money was not his primary motive in pushing toward an autonomous space station. When you head for the planets, he said, "you have to be autonomous; you cannot rely on the Mission Control Center for controlling."[30]

With the space station, Carlisle added, "we've come around to a complete reversal. Now we want to put the simplest thing on the ground, and put the most sophisticated thing in orbit." Herman and Carlisle knew that they could not put "the most sophisticated thing in orbit" on the first try. The first elements of the space station, Herman acknowledged, had to be designed "for a cost ceiling." The trick, in his view, was to get NASA to start with a small but sophisticated space station to which the engineers could easily attach technologies of the future as the station grew.[31]

NASA could build a space station that worked like Skylab or the Soviets' Salyut series and make it fly. Herman understood, however, that it would be difficult to plug new technology into an old space station. The Soviets had solved this problem by junking their stations with new models every few years. Herman rejected this philosophy. He wanted to start with a station that would lay the groundwork for the future. He knew that the facility would start small and grow incrementally. Starting small, however, did not mean starting simple. People who were otherwise opposed to one or another version of the space station supported high technology. High technology was a unifying issue; quantum leaps were part of the American culture. Anyway, spending money to study technology did not imply the same kind of commitment as actually building the station. Technology studies were a lot cheaper. The political consensus supporting quantum leaps allowed NASA officials to take a fresh look at space station technology, an approach they could not execute for the whole space station program, where incrementalism was a more powerful constraint.

In the summer of 1982 the Space Station Technology Steering Committee presented Jim Beggs and Hans Mark with a list of technology research programs and the justification for spending more money on them. It essentially concluded the work of the steering committee. The meeting took place as Beggs prepared to put his case for the space station into the White House budget review machine.

13
Budget Wars

Following their June 1982 meeting, the leaders of the Space Station Technology Steering Committee met with Jim Beggs and laid out the conclusions the group had reached. Of the $107 million that NASA planned to spend on space technology in the upcoming year, $30 million supported the space station. That sum needed to be refocused and rescheduled, but it was supportive. Another $30 million, they told Beggs, should be added to the next year's budget to fill in the gaps. Beggs listened to their rationale and put the $30 million in the budget request NASA was preparing for fiscal year 1984.[1]

The leaders of the Space Station Task Force told Beggs that NASA would be ready to start its "phase B" activities on the space station when the following fiscal year began, on October 1, 1983. The working groups were completing their assignments; the "phase A" concept activities were nearing an end. By late 1983, agency employees would be ready to begin the preliminary design work on the space station. They could undertake detailed cost studies. They would be prepared to start the advanced development program that would convert the technology research into machinery that could be installed on the station. And they could start work on their "trade studies," trade-offs that would have to be made in designing the facility.[2] These steps would require a $30-million appropriation, they told Beggs.

Thirty million dollars for new technology, $30 million for planning and advanced development, plus $3 million for work on the science payloads the shuttle would carry to the station—that added up to $63 million. Beggs approved it all and told the people in the comptroller's office to submit the request to the budget examiners in the White House Office of Management and Budget (OMB), the first step in NASA's incremental budget strategy.[3]

Incremental budgeting begins when an agency head initiates an important new program with a small increment laid over last year's base. The agency head tries to downplay the significance of the increment; politicians reviewing the budget focus on the increment and ignore the base. In transmitting NASA's request to Budget Director David Stockman,

Beggs argued that the limited spending activities meant that an official decision to start building the space station would not be required for "about two years." Proposed spending would merely allow NASA "to explore the requirements, operational concepts, and technology associated with permanent space facilities," as President Reagan had directed in his July 4 speech. Approval of the $63 million, however small a slice in NASA's $7.3-billion budget request it seemed, would nonetheless imply a commitment to proceed. Beggs was prepared to appeal the issue all the way to the President to get that commitment.[4]

The budget examiners at OMB insisted that NASA officials reveal how much they wanted to spend the following year as well. Politicians tended to treat these as fantasy funds, not the basis for real decisions, since Congress appropriated dollars for only one fiscal year at a time and viewed the multi-year projections as unrealistic anyway. The President's budget examiners asked, nonetheless. Beggs told them. Sixty-three million dollars for the space station for fiscal year 1984, $123 million for fiscal year 1985.[5]

As official Washington got back to work after the August 1982 congressional recess, Beggs sent his program managers over to the Executive Office of the President to explain NASA's request for funds. John Hodge went for the Task Force, taking thirty viewgraphs with him instead of the usual twenty. Meeting with the examiners who reviewed NASA's budget, Hodge explained the activities of the Task Force, the Technology Steering Committee, and the various working groups. He described the twelve boundary conditions, emphasizing NASA's commitment to define requirements before proposing designs. He explained the international interest. He talked about the mission analysis studies and the importance of industrial involvement. He laid out the budget in detail and told the examiners how the money would be spent.

Hodge told the budget analysts practically everything except what they wanted to hear. He did not lay out the alternatives to a space station or explain why a $4- to $6-billion space station should take priority over other government programs. Instead, Hodge inserted a single sheet titled "Why a Space Station May Be Essential." The space station, he reasoned, would maintain U.S. leadership in space, enhance national security, stimulate commercial activity in space, promote space science, stimulate technology, enhance national prestige, and keep the NASA space program alive. It would establish a permanent presence in space, free-up the shuttle, service satellites, make space construction possible, and bring deep space closer to earth.[6]

The budget examiners listened to Hodge and the other NASA executives, analyzed the requests, and went off to consult with the people who ran the OMB. The largest office in the galaxy of presidential agencies, the Office of Management and Budget employed 596 of the 1,600 persons who labored in the Executive Office of the President.[7] Political appointees ran the office; career civil servants staffed it. To oversee all science, energy, and natural resource spending, including NASA's budget, Stockman had appointed Frederick Khedouri, who had served as Stockman's legislative director when Stockman served in the U.S. Congress. The budget examiners worked closely with Khedouri in preparing their response, which they delivered to NASA two days before Thanksgiving, on November 23, 1982.

Not a nickel for the space station, they said—not this year, not next year. They would recommend that NASA not be allowed to spend any new funds on the space station for at least two years.

Six days later Beggs filed a "reclama," a term taken from the Latin verb meaning to cry out or protest loudly. An official response to the budget examiners first "mark," the "reclama" gave agency heads a chance to propose a smaller figure in their search for something at all. Instead of $63 million for fiscal year 1984, Beggs said he would take $53 million. But he wanted approval for a $128-million recommendation in fiscal year 1985 instead of his earlier request of $123 million.[8]

The budget officers told Beggs the following story: They knew that NASA was already spending $54 million on the space station that year. Agency officials had admitted that $29 million of their generic technology program was applicable to the space station. The budget officers knew that NASA executives had put at least 250 of their employees to work full time on space station planning. NASA executives had established a Space Station Task Force with full-time employees working in Washington, D.C., and NASA center directors had assigned employees to work solely on space station studies in the field. At $60 thousand per person per year, that totaled $15 million. The budget officers had uncovered another $10 million in space station and unmanned platform studies. NASA was asking for $53 million, they said, but the agency was already spending $54 million. Thus, the budget officials told Beggs, they would recommend no new funds, nothing. Furthermore, they said they would order NASA to terminate some of its contracts with the aerospace companies that were helping to develop NASA's space station plans.[9]

The leaders of the Space Station Task Force delivered the following arguments to the NASA officials conducting the budget negotiations with OMB. The agency needed $53 million in fiscal year 1984, Daniel Herman argued, to conduct "the major technology trade studies that are crucial in the definition of a technology baseline." The money would allow future users of the space station to define user needs and allow NASA to translate those needs into requirements for the facility. "A reduction to $35 million," Herman explained, "can be accommodated by delaying the technology trade studies by six months." This would delay the launch of the space station by one to two years and perhaps have the effect of "restricting the technology options and hence either increasing risk or restricting growth potential," he warned. "However, the program is still a viable continuing effort at this level," Herman said. "Below $30 million, we cannot continue both the systems definition and technology trades in FY 84. It makes no sense to do either effort without the other."

The other members of the Task Force argued just as vigorously. Any appropriation below $35 million, Terry Finn maintained, "destroys Beggs/NASA credibility." It "wastes money." It "tells Europe and Japan we're not serious." It would destroy the "internal NASA coalition on space station planning" and give "permanent presence to the Soviets, by default." If OMB cut the FY 1984 budget below $30 million, Herman concluded, he would "recommend that we terminate the space station activity and re-initiate it at a later date."[10]

Beggs filed a second reclama on December 8, 1982. He would accept $30 million in new funds for fiscal year 1985 and a $40-million base for the following year. If the budget analysts continued to zero out the space station, he would take his case to the President.[11]

The space station was not the most important increment on NASA's platter of new starts. Nor at $30 million was it the largest. In his letter transmitting the agency's 1984 fiscal year request to Budget Director David Stockman, Beggs assigned a higher priority to the construction of one more space shuttle, which would bring the fleet of orbiting spacecraft up to five. NASA officials badly wanted the fifth orbiter. A fifth orbiter, Beggs argued, would allow NASA to sell launch services to business firms and foreign users and permit the agency to stick with its proposed flight schedule if anything happened to one of the other four. To get started on the new spacecraft, Beggs asked Stockman to approve $200 million to be spent in fiscal year 1984.[12]

Beggs had carefully prepared the way for White House approval of the fifth orbiter. He had won the support of the Air Force, which had already invested heavily in the development of the space transportation system and felt that NASA needed a fifth spacecraft in order to meet the flight schedule for launching military satellites. Beggs had won an endorsement from the special group on the National Security Council that drafted White House space policy. He had the support of the aerospace industry that would build the orbiter in the President's home state. NASA's position on the fifth orbiter was much stronger than its stand on the space station, but then Beggs was asking for four times as much money for the orbiter than for the space station in the FY 1984 budget.[13]

Beggs could probably appeal one new initiative to the President if the budget analysts turned him down. It would be very hard to appeal both the space station and the new orbiter in a tight money year.

Fred Khedouri, Stockman's principal assistant on these matters, offered Beggs a compromise. The budget office would not zero out the space station. He would recommend that NASA be allowed to spend $14 million in fiscal year 1984 and give tentative approval to another $14 million for fiscal year 1985.[14] "Everybody was crushed," said Clairborne Hicks, a member of the Space Station Task Force. "OMB came in and said, 'There isn't any way that we are going to let you start phase B,'" the next step in the development of the facility. "They were the dark days," Hicks said, remembering the budget wars at the end of that year.[15]

Beggs knew all along that the President's top budget officers would oppose the space station. They had been hired to oppose big spending and new programs. Beggs knew that in one way or another he would have to appeal the issue to the President. To appeal on the space station that year, he would first have to appeal to the Director of the Office of Management and Budget and the budget review council that advised him.[16] Representatives from the White House staff, the cabinet, and sometimes the Treasury Department would sit in on the budget appeal. "They would really work you over," said one official who had participated in the process. The review council would remind Beggs of the need to be a team player and not rock the boat.[17] Khedouri knew that Beggs would lose an appeal to Stockman during the "Director's Review." David Stockman, the White House budget director, had not received any mandate from the President regarding the space station. Without such a mandate, Stockman was under no obligation to approve a program he personally opposed, especially when NASA was already spending more money on the space station than it was asking for.

Beggs, of course, could appeal Stockman's denial to President Reagan. Beggs had just met with one of the President's closest advisers, Edwin Meese, to test White House support for NASA's initiatives. Earlier that year, one of the agency's closest allies on Capitol Hill, Senator Harrison Schmitt, a former astronaut and chairman of the Senate subcommittee that authorized NASA programs, had met with President Reagan to test the President's sentiments on new space initiatives.[18]

"He was a famous astronaut who had walked on the moon," Stockman complained, "and he had never quite gotten over it." Schmitt came away from the meeting without any new commitments.[19]

Beggs decided not to appeal. "After more mature thought and doing a little more checking, I didn't think the timing was right. So the OMB offered me a compromise, saying okay, they wouldn't kill it completely, they would allow me to spend a little money on it."[20] Beggs took OMB's offer of $14 million. Outwardly, he and the leaders of the Space Station Task Force portrayed it as a compromise. In fact, it was a defeat. It sent a clear signal to NASA's leaders, including the people running the Space Station Task Force, that their incremental budget strategy had failed. They would not be able to get the space station proposal approved by pushing a small piece of it through the White House budget office. The episode gave the NASA leadership its first clear picture of the depth of the opposition to the space station among members of the President's cabinet and his White House staff. It revealed the degree to which NASA's leaders were willing to let the scope of the program be whittled down before making an official appeal. In all, it was a sobering experience.

NASA got the $14 million, which it spent in fiscal year 1984. The Budget Director, the President, and Congress approved the item without giving it much attention. NASA continued to direct other funds from its technology and space flight programs into space station planning, as agency executives had done for more than twenty years without ever receiving approval to go ahead and build the facility.

As for the fifth orbiter, the leadership at OMB offered Beggs another compromise. They would not let NASA spend $200 million to build a new spacecraft, but they would allow the agency to consume $100 million buying spare parts for the other four.[21] Following the publication of the President's budget, a reporter asked Beggs whether he considered the decision against the fifth orbiter "a lost new start." "Yes," said Beggs, "it's been widely publicized that we asked for the fifth orbiter." Why didn't he appeal the decision to the President? "As we got into the examination of the rationale," Beggs explained, we "could not sustain an argument for it." Beggs would not move forward to the President without an airtight rationale.[22] And the space station initiative? "I decided to keep my powder dry and flush this thing up in a different way."[23]

In 1952, Wernher von Braun urged the United States to build an earth-orbiting space station. Illustrated in Collier's magazine, the 250-foot-wide wheel would rotate so as to produce artificial gravity.

Von Braun suggested that the United States deploy the space station by inflating it. By 1961, NASA officials had built models of inflatable space stations—on the ground.

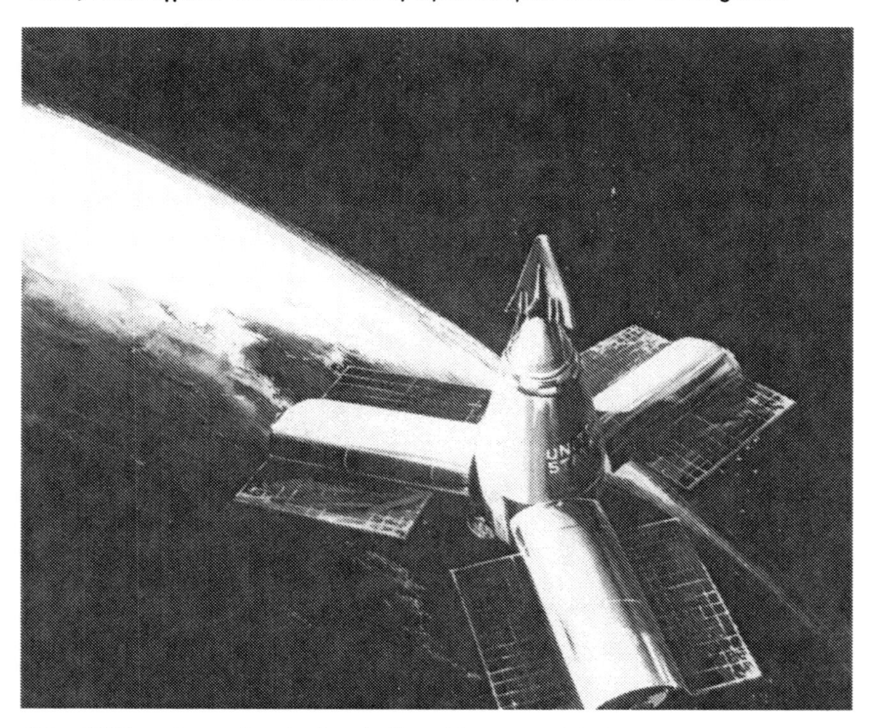

Other NASA engineers thought that the United States could deploy a large rotating space station by launching it folded up. Once in orbit, the space station would open like an umbrella.

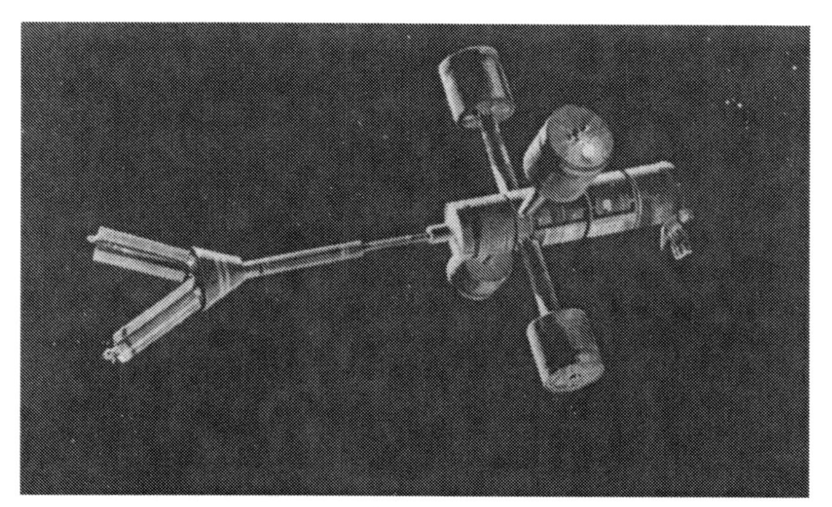

Following the landing on the moon, NASA officials sought approval for a large pinwheel-shaped space base as one element in a long-range plan that also included a space transportation system, a lunar base, and a mission to Mars. When the White House rejected the long-range plan, NASA officials adopted an incremental strategy. They would build the space shuttle first, then ask for a space station.

By 1971, NASA officials realized that any U.S. space station would have to be built out of modules, delivered by the space shuttle and fastened together in orbit. The space station would not rotate, but would utilize the advantages of weightlessness for space science and manufacturing.

In spite of all their planning, NASA officials received no authority to start work on a large, permanently occupied space station during the 1970s. In 1973, they were allowed to launch Skylab, the prototype for a future space station. Three teams of astronauts spent a total of 171 days in the orbiting workshop. Deserted, the workshop in space fell to earth in 1979.

Beginning in 1971, the Soviet Union launched a series of small experimental space stations called Salyut. this is Salyut 7, with a Soyuz spacecraft docked at the rear port on the lower side.

*Even though no space station had been approved, NASA employees continued to develop plans for one. The designs moved away from the pinwheel-shaped space stations of the early 1970s toward structures held in place by latticework. This plan for a Space Operations Center, issued in 1981 by officials at the Johnson Space Center, included a space hangar (*lower left*) where astronauts could repair satellites before launching them into higher orbits.*

Scientists and engineers at NASA's Marshall Space Flight Center proposed a more modest "space platform." This low-cost space station contained open pallets for space experiments (center and lower right) *and a small science lab* (center left)*. Like other space stations under study, the space platform contained two large solar panels for generating electric power, a radiator fin for dissipating heat, a logistics module containing supplies* (center top)*, and a habitation module for the crew* (center right).

NASA Administrator James Beggs, appointed to head the space agency in 1981, announced that he would seek approval for a space station once the Space Transportation System became operational. The successful test flights of the space shuttle lay the groundwork for what Beggs called "the next logical step."

In May 1982, Beggs deployed four NASA employees to organize a special Space Station Task Force. More than five hundred NASA employees eventually became involved in the effort to plan and promote the space station, even though the President had not yet approved the program.

Top left to bottom right, *Task Force leaders John Hodge, Robert Freitag, Terence Finn, and Daniel Herman.*

President Ronald Reagan sought ways to develop a more aggressive space program after he took office in 1981. NASA officials invited him to Edwards Air Force Base to welcome back the crew from the final test flight of the space shuttle Columbia on July 4, 1982, and to announce his new space policy.

In spite of the success of the shuttle test flights, however, President Reagan did not embrace the space station initiative at that time. Speaking to a crowd at the July 4 shuttle landing, he made a vague commitment to "a more permanent presence in space" instead.

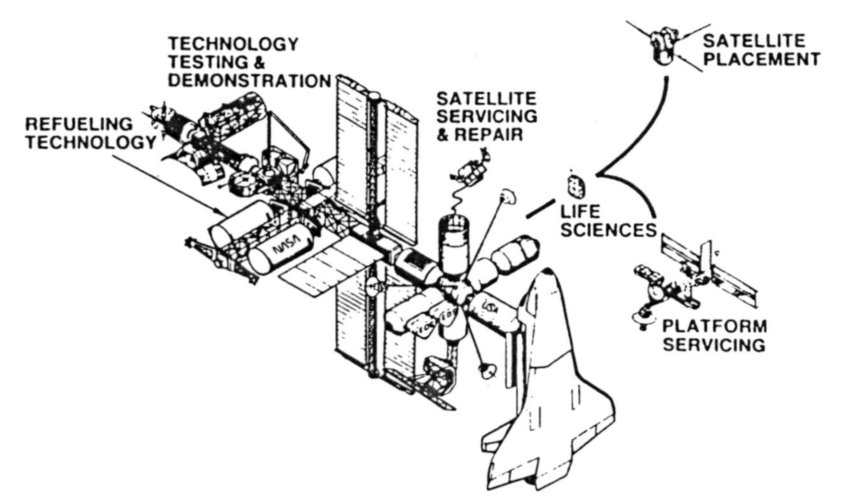

Leaders of NASA's Space Station Task Force refused to issue a specific design for the proposed space station, concentrating instead on the missions the space station would perform. The first rough sketch of the Task Force space station did not appear until mid-1983, on one of the viewgraphs prepared for the briefings the Task Force leaders frequently gave.

Presidential assistant Gil Rye inspects the model that NASA officials prepared for President Reagan as he reviewed the space station proposal. Rye organized the White House interagency group that studied the initiative for the President.

On December 1, 1983, President Reagan met with members of his Cabinet Council on Commerce and Trade to hear their advice on the space station initiative. Administration officials were deeply divided over the issue.

In his January 1984 address on the State of the Union, President Reagan directed NASA "to develop a permanently manned space station" and to do so by 1994. Congress still had to approve and fund the program.

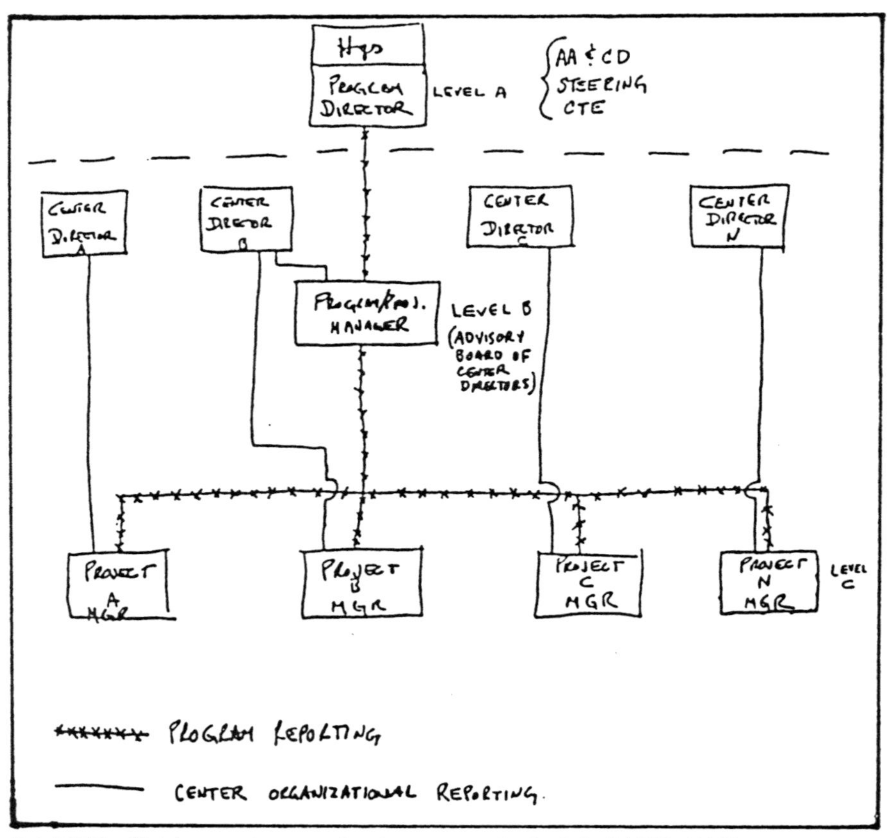

Top NASA officials started to organize the program. In the fall of 1983, NASA center directors produced a "barbed wire" organization chart that promised cooperation in exchange for little supervision. Barbed wire (the hatched lines) rather than a strong program office bound together NASA field centers working to design and build the space station. NASA Administrator James Beggs approved this "lead center" approach in February 1984.

In May 1984, at the London Economic Summit, President Reagan followed up on his invitation to the leaders of Western Europe, Canada, and Japan to join with the United States in building an international space station. International cooperation helped win support for the program within the U.S. Congress.

By 1987, the planned space station had evolved into a graceful 500-foot-long design. Hemmed in by budget deficits and their incremental approach to space policy, NASA employees continued to reshape the space station as politicians reconsidered the program.

PART III

14
Positions

NASA officials had tried to get the space station approved many different ways and had failed each time. They had failed to get the space station approved in 1969 by presenting it as part of their long-range plan. They had failed to entice President Reagan into making a quick, Apollo-style commitment. They had failed to sneak it through as an incremental wedge in the agency's budget. By the start of 1983, NASA executives recognized that they would not be able to get the program approved without submitting it for some sort of comprehensive White House review.

Three parties would be deeply involved in any such review. David Stockman, as Director of the White House Office of Management and Budget, wanted to review it for conformance to the President's budget-cutting plans. George Keyworth, the President's science adviser and head of the White House Office of Science and Technology Policy, wanted to analyze the scientific merits of the proposal. Members of the defense community, through their representatives on the National Security Council and its staff, wanted to examine the effect of such a major new initiative on their equally ambitious space plans.

"I was as much of a space buff as the next person," said Stockman. His fascination with extraterrestrial exploration did not make him a space station advocate, however. "Space stations are high-tech socialism," Stockman announced, in his typically ideological way. "Big spending to me, sound investment to NASA administrators."[1]

At the age of thirty-four, upon his appointment as the President's budget director, David Stockman had established a reputation as one of the brightest young minds in the so-called Reagan revolution. The eldest of five children from a family of Michigan fruit farmers, Stockman had progressed from Michigan State University to Harvard Divinity School to the staff of the U.S. House of Representatives, in the process embracing no less than four different ideological viewpoints. He had been a Goldwater conservative, a neo-Marxist, a follower of Reinhold Niebuhr, and a progressive Republican. By 1975, his search for a coherent political philosophy had led him back to the conservative faith. His search for political influence had led him, in 1976, to win election from the Fourth

Congressional District of Michigan, committing in the process the cardinal sin for a young congressional staffer of running against an incumbent representative from his own political party. Congressman Stockman had developed a reputation as "a walking computer," a repository of knowledge on economic policy and federal spending. After two terms in Congress, his conservative politics and some well-placed political backers won for him an appointment as President Reagan's keeper of the purse.[2]

Stockman's attitude toward NASA flowed freely from his opinion of government handouts in general. Before his election to the House of Representatives, Stockman had published the lead article in *The Public Interest,* an academic journal that served as a forum for the new conservative movement. Developing a thesis that would guide him in later years, Stockman argued that American politics corrupted government programs designed for presumably useful ends. He took aim at government programs designed to help the poor. "A great social pork barrel," he called them. If the $80 billion in annual social welfare spending actually went to those in need, he argued, "the officially estimated poverty gap of $12 billion could be closed seven times over." Instead, he said, the money went to fuel welfare bureaucracies and professional groups that ministered to the poor, diluting the funds in a effort to build a political constituency for the anti-poverty programs among the voters at large.[3]

Stockman extended his "pork barrel" thesis to other programs as well. "Institutional forces in the military," he announced in one outburst of intellectual impatience, "are more concerned about protecting their retirement benefits than they are about protecting the security of the American people."[4]

And Stockman doubted the good intentions of the leaders of the space program. NASA's budget, he wrote in a manifesto to Ronald Reagan before the inauguration, could be cut by one-third without harming the nation's leadership in space. To a protesting senator who argued that the nation needed a large space program to retain its technological superiority, Stockman replied that the way to spur technological progress "was to reward private inventors, entrepreneurs, and investors with lower taxes."[5]

In his skepticism about the aims of the space program, Stockman found himself allied with a number of liberal senators and congressmen. "I am concerned," Wisconsin senator William Proxmire told NASA executives, when confronted with the space station initiative, "that it will proceed regardless of the real need for such a program because your agency needs it more than the country needs it." Criticizing NASA's approach to the initiative, Proxmire announced: "I have long believed that your agency has a strong bias toward huge, very expensive projects because they keep your centers open and your people employed."[6]

As he left office, President Jimmy Carter asked Congress to raise NASA's budget by $1.2 billion, a 22 percent increase. As he came in, David Stockman convinced Ronald Reagan to cut that increase back to $600 million. More important, Stockman laid out a five-year plan that leveled NASA's spending for two years and then turned agency spending down as the space transportation system went operational. Stockman's budget plans for NASA contained no room for new initiatives, neither space stations nor other alternatives. When Senator Harrison Schmitt warned Stockman that Europe and Japan would gain ground on the U.S. economy as a result of the cutbacks in space technology, Stockman responded: "I was not aware that either the French or the Japanese had yet landed on the moon."[7]

Stockman's skepticism about the purpose of the space station was shared by the President's chief science adviser, George Keyworth, who sat one floor above him in the Old Executive Office Building. Like Stockman, Keyworth projected a boyish look underneath a tuft of black hair and a pair of thin-rimmed glasses. Like Stockman, Keyworth was very bright. Stockman had studied philosophy and American history; Keyworth had a Ph.D. in physics. Stockman had attached himself to a crew of intellectual congressmen in his rise to the top; Keyworth was the protégé of a group of conservative scientists, including Dr. Edward Teller, the so-called father of the hydrogen bomb and a hard-biting, apocalyptic political adviser.

"I have asked ever since I came to this office for people to tell me what we will do with the space station," Keyworth explained. "Neither NASA nor the space community has been able to define what those men would do who would be up on that space station."

"The main reason people would like to build a space station is to put man in space," Keyworth offered. "Why do we want to do that when we are just entering the era of robots and automation? The Soviet Union is way behind us in microelectronics and robotics and all these automation technologies, which is why they have put men in space. Why should we put men in space when our industrial base has this enormous capability to do tasks without men?"[8]

"Don't emulate an inferior technology," he told members of the House Appropriations subcommittee that would have to fund the space station. "For the nation that is the world's leader in automation technologies and in microprocessors and the underlying technologies, to take a step backwards to emphasize man in space at the very time when we are best equipped to use our automation technology," he said, "is a most unfortunate step."[9]

In his skepticism, the forty-one-year-old Keyworth represented people who shared a very different vision of the exploration of space than the one that dominated NASA. The United States had just completed a decade of space flight in which some of the most spectacular voyages were made by machines, from the Viking spacecraft that landed on Mars in 1976 to the Voyager twins that initiated their mission to the outer planets with the 1979 Jupiter flyby.

These people doubted the wisdom of sending humans into space to do jobs that could be done faster and more accurately by machines. "More than two-thirds of NASA's budget is devoted to the development and operation of the space shuttle fleet and, prospectively, to a huge permanently manned space station," physicist James Van Allen claimed. Space scientists like Van Allen complained that development of the shuttle had resulted in the "slaughter of the innocent," cutbacks and cancellations affecting "dozens of programs" for the unmanned exploration of space. Van Allen feared that the space station would delay future projects as well. "The progressive loss of U.S. leadership in space science can be attributed, I believe, largely to our excessive emphasis on manned space flight."[10]

Keyworth challenged the supporters of the space station to provide a convincing rationale for the multi-billion dollar facility. Writing to the lawmakers on Capitol Hill, he asked for assurance that the space station "is the right step for continuation of the nation's space program; that our other on-going programs are not adversely affected by a new start of this magnitude, by this I mean the objectives of a fully operational cost-effective shuttle and continuation of the planetary and sensor developments underway at NASA"; and finally that "requirements from the civil and national security communities justify the major resource expenditures" such an initiative would require.[11]

Stockman's and Keyworth's words echoed the past. When President Kennedy had wanted to send American astronauts to the moon, his science adviser had told him that the venture could not be justified "on scientific grounds." Kennedy's budget director had argued that just as much could be learned "by sending instruments out instead of men," and at a fraction of the cost of a manned probe. President Kennedy decided to send astronauts anyway. "It didn't impress him very much that the scientists said you can do it with instruments," the budget director remembered. "If there was something to be learned, I'm sure he felt in the seat of his pants that it would make a difference to have men out there." And so did most NASA officials, who had grown up in the age of human flight.[12]

Keyworth headed a tiny division in the Executive Office of the President. The White House Chief of Staff allowed him to hire eleven full-time permanent employees, less than one-tenth of one percent of the total number of people who worked for the President. Created in 1962, the office that Keyworth headed had so little influence that President Nixon abolished it in 1973. (Congress reinstated it in 1976.) Keyworth, moreover, had no experience in the high-altitude world of Washington politics. He had spent his entire career as a research physicist at the Los Alamos National Laboratory in the mountains of New Mexico.[13]

David Stockman, on the other hand, had ten years' experience with political infighting in Washington. He directed the oldest and largest institution in the Executive Office of the President. "OMB," said Stockman, referring to the 600-person office he headed, "is the needle's eye through which all policy must pass." He knew how to mobilize support to win battles among members of the White House staff, and he was not reluctant to use his allies on Capitol Hill to put pressure on the President. Nonetheless, Stockman was not a member of President Reagan's White House inner circle. He did not have direct access to the President and was viewed by many experienced politicians as something of an ideological gadfly.[14]

From previous experience, NASA officials knew that the opposition of the President's budget director and science adviser could be overcome. The third party to the White House review, however, created an entirely different set of problems. Never in the twenty-three-year history of the agency had NASA officials launched a major manned space flight initiative over the opposition of top officials in the Department of Defense.

The 1961 staff paper setting up the flight to the moon had been drafted in the Pentagon by a group of officials headed by NASA Administrator James Webb and Defense Secretary Robert McNamara, both of whom understood the military significance of a strong space program. The Saturn V rocket that took American astronauts to the moon was designed by employees of the Department of the Army working at a military installation placed under the jurisdiction of the National Aeronautics and Space Administration by the President of the United States.[15]

When NASA officials sought approval for the Space Transportation System in 1969, the Secretary of the Air Force wrote a letter endorsing the concept and sent it to the White House group reviewing the proposal. Eight years later the Deputy Secretary of Defense agreed that the military would "use the STS as the primary vehicle for placing payloads in orbit," thereby elevating NASA's spaceship to a status higher than that of the rockets the military controlled. The Secretary of Defense coughed up, by one estimate, nearly $4 billion to help build the Space Transportation System, a sizable contribution to the funds NASA needed to launch the program. Much of the Defense Department contribution went to build a new spaceport on the West Coast of the United States, from which NASA and the Department of Defense planned to send the shuttle into polar orbits. When NASA ran into cost problems, top Defense Department officials agreed that the department "should pay a fair share price to have payloads placed in orbit by the Space Transportation System," thereby assuring NASA of a reliable stream of revenues to operate the system. When NASA needed more money to cover cost overruns on the program, the Defense Department rolled out its big guns to lobby the White House and Congress to approve NASA's requests.[16]

When Jim Beggs took over as NASA Administrator in 1981, the White House sent in the previous Secretary of the Air Force to be his deputy. As Undersecretary and then as Secretary of the Air Force from 1977 until 1981, Hans Mark had been a strong advocate of space exploration and an early supporter of the space station. The official Defense Department position on its need for a space station, however, remained ambiguous at best.[17] Fifteen years earlier, Defense Department officials had attempted to launch their own space station. A previous Secretary of the Air Force and Chairman of the Joint Chiefs of Staff enthusiastically embraced what they called the "manned orbiting laboratory." The Air Force set up a special office to develop the program and hired six prime contractors to design and build the facility. The Air Force selected seventeen astronauts to operate the station and pilot a military spacecraft to and from the station. It was an open secret that the orbiting laboratory would be used primarily for military reconnaissance and surveillance, a man-tended spy station flying high above China and the Soviet Union.[18]

In explaining the decision to scrap the program in 1969, Deputy Secretary of Defense David Packard cited "advances in automated techniques for unmanned satellite systems." Friends of the defense establishment suggested that the decision was rammed down the throats of the professional military by high-ranking officials in the White House. "President By-passed DOD & Joint Chiefs to Cancel MOL," the headline writers at *Space Daily* complained. Reporters fingered White House Assistant Budget Director James Schlesinger (later to be appointed Secretary of Defense) as the "MOL assassin."[19]

NASA officials like John Hodge interpreted the cancellation of the DOD space station in somewhat different terms. They believed that the decision to cancel the program originated within the Air Force itself, not in the White House. Although the decision technically came down from the White House, officials like Hodge saw the decision as a victory for what they called the "black Air Force" (the intelligence-gathering arm of the service) over the "blue Air Force" (the pilots and rocketeers).[20] Officers in the "black Air Force," along with their counterparts in the Central Intelligence Agency, developed the highly sophisticated instruments that took pictures of troop movements and missile tests from space and sent those images back to the ground. This faction within the Air Force argued, apparently successfully, that unmanned reconnaissance satellites could do the work of manned observatories at a fraction of the latter's cost, a highly popular argument in the Republican-controlled White House, where budgeteers felt pressure from the Democrat-controlled Congress to put at least one big defense item on the cutting block.[21]

The Defense Department had announced the termination of the Air Force manned space program on June 10, 1969, just as NASA presented its own plans for a space station as a follow-on to the Apollo program. Two months later, the Secretary of the Air Force published the department's official position on NASA's proposed civilian-run space station. "Even though the development of a large manned space station appears to be a logical step leading to further use and understanding of the space environment," the Air Force Secretary wrote, "I do not believe we should commit ourselves to the development of such a space station at this time." Lacking support from the Defense Department and finding few allies in a cost-conscious White House, NASA's 1969 space station proposal died.[22]

Air Force opposition to any military man-in-space program continued through the 1970s. In 1979 the commander of the Air Force Systems Command repeated the company line. "We do not need a series of large military manned space stations," he said. "We know that we can put people into space," he explained, "but to put military people in space just because it's something we can do is not my idea of a judicious expenditure of our scarce resources."[23]

Still, space station supporters saw some promising signs. Shortly after taking over as Secretary of Defense in 1981, Caspar Weinberger hinted that the department might support a space station initiative. "It is, of course, difficult to predict how the future will unfold," he said. But, he added, "permanently manned space stations serving the needs of mankind are developments certainly in the realm of reasonable expectations." Weinberger's chief assistant for research and engineering made a similar statement. "We agree with them," he said of the NASA officials pressing for a space station. "That is the next civil effort that ought to be done." Defense Department officials did not, however, offer full departmental support for the initiative.[24]

NASA officials were determined to win such a full endorsement. "We were trying to get their support at that time," admitted Jim Beggs. The leaders of the Space Station Task Force listed "DOD participation" as one of the thirteen "boundary conditions" defining the program and got the Department of Defense to join the group constructing the list of missions to be performed on the station. NASA Deputy Administrator Hans Mark, plying the contacts he had developed as Secretary of the Air Force in the previous administration, worked to win Defense Department support. To get that support, NASA executives sought to identify some functions the Defense Department could perform on the space station in order to justify military participation.[25]

The anti-man-in-space majority within the Air Force, however, continued to argue that a space station would serve no real military need. "The primary concern of the Air Force people was the development of what they called 'assured access to space,'" said Hans Mark. The Air Force people did not want to see NASA go bounding down the path toward a space station until NASA spent more money perfecting the space shuttle, which the Defense Department had agreed to use as its "primary vehicle for putting payloads in orbit."[26]

Even if Defense Department officials shifted their position and embraced the need for soldiers in space, the defense community still might not endorse NASA's space station. Experience with the space shuttle had taught Defense officials a convincing lesson on the challenge of trying to carry out military missions on a facility run by an agency they could not control. "The fact is," said Hans Mark, "that most of the leaders of space activities in the Air Force were not happy with that role." NASA made its space station proposal following a period when many senior officers in the Pentagon were looking for a way to pull out of the space shuttle program and launch their own satellites with rockets under Air Force control. Support for a civilian space station struck them as one more step in the wrong direction.[27] "We are just very cautious about being too bullish on space stations," Weinberger's assistant for research and engineering explained, "until we really know what it looks like as a program."[28]

Beggs had few allies on Keyworth's staff and fewer still on Stockman's crew. He and his Task Force still needed some way to move the space station initiative through the White House policy review process, one that gave Stockman and Keyworth an opportunity to comment on the proposal but not to gut it as the budgeteers had attempted to do twelve years earlier with the Space Transportation System. Of the three parties with whom NASA had to deal, the leaders of the national security community seemed the most amenable. Given the bureaucratic politics of White House review, Beggs needed a route through the White House which was dominated by military representatives who might listen to NASA's point of view.

15

The White House

Observers of the American presidency tell a time-honored anecdote about President Abraham Lincoln, who, as the story goes, issued a decision in the face of the unanimous opposition of his cabinet. Having called for a vote at a cabinet meeting, Lincoln found himself the sole member in favor of the move. "Seven noes, one aye," he said, announcing the outcome. "The ayes have it."[1]

The anecdote illustrates the point that, technically speaking, the President of the United States suffers no legal obligation to submit important decisions to his principal subordinates for their consent or consultation. Under the American Constitution, within the executive branch of government, the President has the only vote that counts.

The anecdote makes a nice point, but with a fault. It is not true; it did not happen that way. The real story does a much better job of illustrating how presidential decisions are actually made. Only eleven days after his inauguration, President Lincoln polled his cabinet on the question of whether a military convoy should resupply the Union fort on the coast of South Carolina, Fort Sumter. Although Lincoln favored the move, members of his cabinet feared that the Southern states would interpret such a mission as a first step toward civil war. Without an effort to resupply, however, the fort would have to be abandoned to the secessionists. The cabinet, including the Secretaries of State and War, voted six to one against resupply, expressing their willingness to abandon Fort Sumter in the hope of negotiating a settlement with the South. Shaken by the vote, Lincoln reconsidered his position, delaying any final decision as the leaders of the South and the members of the cabinet debated their next move.

Lincoln had appointed to his cabinet the principal rivals within his party for the presidency, all of whom thought themselves better qualified for the job. As they plotted against Lincoln, he sought to manipulate them. Lincoln reframed the issue, advancing the more radical proposition that the federal government surrender both Fort Sumter in South Carolina and Fort Pickens in Florida. The advocates of negotiation, including Secretary of State William Seward, openly argued against the resupply of Fort Sumter. But they could not embrace the surrender of the second fort, since, in the words of one leading historian, "it was common knowledge that Pickens could be held indefinitely." As Lincoln continued to meet with the group, the vote shifted in his favor. With a majority behind him, Lincoln initiated the resupply effort, negotiations failed, and the Civil War began.[2]

The full story illuminates the true nature of presidential government. While the President has no constitutional obligation to poll the persons he has appointed to head the principal departments in his government, presidents regularly consult the members of their cabinets and wait for some consensus to emerge before making important decisions. It is extremely rare for a president to charge off on an important issue without giving the members of his administration the opportunity to present their advice and, where possible, line up behind him.

The nature of that consultation tends to vary with the decision-making style of each president. Lincoln manipulated his cabinet. Reagan delegated decisions to his. Like President Nixon in his approach to the shuttle decision, President Reagan preferred to give his advisers a general commission, wait to see if members of his administration could negotiate away their differences, then step in at the end to resolve any remaining areas of dispute. As the price for putting a new initiative on the desk of a delegating president, agency heads like Jim Beggs were expected to resolve as many details as possible through the members of the President's staff before taking the issue to the chief executive.

The process for resolving such details had grown more formal as the executive branch had grown in size. To reach President Lincoln, members of his cabinet entered the White House from Pennsylvania Avenue and climbed the grand staircase to the second floor of the executive mansion, often passing hoards of office seekers waiting to see the President or one of his two secretaries. The President's office and the Cabinet Room sat at the top of the stairs, overlooking the south lawn of the White House and the Treasury Department to the east.

In Lincoln's time, the cabinet consisted of seven men appointed by the President to run the seven major departments of government. One hundred twenty years later, President Ronald Reagan, with the help of his aides, selected thirteen men and women to head the major executive departments, appointed another ten to manage the major independent agencies, chose ninety special assistants to help him run the White House, backed them up with a staff of 1,500 executive office employees, and nominated, as the positions arose, nearly two hundred officials to run the major government corporations and regulatory commissions, over which he possessed varying degrees of control.

Officially, the modern cabinet consisted of the heads of the thirteen executive departments, plus "certain other executive branch officials to whom the President accords Cabinet level rank," such as the U.S. Ambassador to the United Nations. Then the President would let in heads of independent agencies (such as the NASA Administrator), who were "invited from time to time for discussion of particular subjects," whereafter members of the White House staff insisted on attending during discussions in their areas of concern. Then the Vice President wanted to come, and the cabinet needed a full-time secretary, who needed a full-time staff.[3]

Faced with a hall full of supplicants looking for seats in a small-sized room, modern presidents no longer convened the cabinet as Lincoln had, twice weekly for counsel. In the place of the single cabinet, modern presidents and their chief assistants created various special councils and interagency groups through which policy passed on its way to the presidential suite. Only those department or agency heads affected by the relevant policy, along with the appropriate members of the White House staff, sat on any particular council. Councils appeared and disappeared; membership shifted. Unlike the practice in Congress, where the jurisdiction of standing committees remained intact from election to election, each new president felt obliged to reshuffle the membership of the White House councils to erase the institutional biases created by his predecessor. The large number of councils, combined with their shifting membership, gave a special advantage to top policy makers skilled enough to manipulate the process. By guiding an issue into the right council, an official might predetermine the outcome of the review.

Meetings of the special councils did not take place exclusively in the Cabinet Room, completed in 1909 next to the site of the President's Oval Office. Conceived during the administration of Theodore Roosevelt, the plan for the first floor of the West Wing of the White House originally called for six large offices, a conference room, a comfortable vestibule, and a spacious lobby—in addition to the quarters of the chief executive and the Cabinet Room. Ever anxious to crowd closer to the center of power, the burgeoning White House staff carved up the airy design to create twenty-one tiny offices, some no larger than the bathroom in a private home. The lobby shrunk to less than one-fifth of its original size, sacrificing splendor for office space.[4]

Jim Beggs entered the West Wing through a basement door adjoining an alleyway on the west side of the West Wing. The alleyway separated the president's working space from the magnificent old Executive Office Building, a monument to creative architecture in a city full of undistinguished office buildings. Beneath its numerous chimneys and Second Empire façade, the building housed the expanding presidential staff, which spilled out of the West Wing, through the Executive Office Building, and across Pennsylvania Avenue. The Departments of State, Navy, and War, for whom the building was originally constructed, had long since left for more-ample quarters closer to the Potomac River.

Beggs worked his way past the White House guards to a small, windowless room in the basement of the West Wing. Tulips and grape hyacinths bloomed in the rose garden next to the Cabinet Room, but Beggs did not see them that morning. Like many other presidential briefings that involved national security issues, this one on April 7, 1983, took place in the more secure quarters of the Situation Room, an undistinguished chamber that served as the White House command and control center. The space station concept that Beggs discussed with the President and his staff that morning had surfaced during meetings scheduled to brief Reagan on the activities of the Soviet Union in space. "We tried to get him briefed on what the Soviets were doing," Beggs explained, "with particular emphasis on the fact that the Soviets were flying a space station."[5]

Although that was how the space station issue officially came up, Beggs had not emphasized the national security implications of the space race in his pitch for presidential support. Beggs presented NASA's plans for the future so as to emphasize national leadership in commerce and technology. He had told Reagan about "the importance of setting new goals in this program and the advantages of those: in motivating the country and getting it back again, pointing into the future, the emphasis we gave to research and development and so forth and so on, all the arguments; the halo effect that this program tends to have on R&D, the technological spin-offs that the agency provides to industry in general, the way in which the program motivates young people to aspire to build larger and more important things, the science and technological and the commercial [angle] which he liked very much."[6]

Beggs made no formal presentation to the President at the April 7 briefing, the first one to deal officially with NASA's space station proposal. Reagan did ask about a possible mission to Mars, a question Beggs suspected had been planted by the White House science adviser. Beggs replied that the space station was the "way station" to any such future missions. "You have to have it if you are going to Mars," Beggs explained. "First of all you have got to understand the implications of long-duration activities in space. You have got to develop a lot of technologies that would permit you to take long-duration flights, and when you get ready to do it you have to have a place to stage from. And I went back and recounted the fact that in the original planning for Apollo we would have liked to have gone earth orbit to the moon, but we didn't have time, maybe even didn't have the money, but we certainly didn't have the time." Beggs sensed that interest in a Mars expedition or future missions to the moon would evaporate as soon as the White House priced them out, so he stuck to the short-term goal of the space station. "Generally," Beggs thought, "the President bought those arguments."[7]

Four days later, President Reagan signed a directive setting up a White House staff study on the space station issue. From the directory of White House interagency panels, he and his advisers selected a relatively new council to conduct the study "to establish the basis for an Administration decision on whether or not to proceed with the NASA development of a permanently based, manned space station." The presidential directive, which the White House issued on April 11, 1983, placed the issue before the Senior Interagency Group for Space, a special council whose membership the President had personally approved the night before his July 4, 1982, speech at the landing of the space shuttle Columbia.[8]

Known around town as SIG (Space), the council had been conceived by George Keyworth, the President's science adviser, architect of Reagan's 1982 space policy, and a well-publicized skeptic on the space station initiative. The council served, as stated in the usual White House bureaucratese, "to provide a forum to all Federal agencies for their policy views, to review and advise on proposed changes to national space policy." As the nation had moved into space, so had the number of departments and agencies with interests in the ether: first the Army, Air Force, and Navy, then the Central Intelligence Agency (for intelligence gathering), then the Arms Control and Disarmament Agency (for treaty verification), and most recently the Department of Commerce (for everything from weather satellites to manufacturing in space).[9] In the original document setting up the coordinating council, Keyworth had written himself in as chairman of the group.[10] That left NASA in a very difficult situation. Space station policy would have to pass through a coordinating body headed by the science adviser; the budget would have to pass through David Stockman, who, like Keyworth, viewed the initiative suspiciously.

An aide to the National Security Council, however, intercepted the document before it reached the President. The White House staffer who snatched Keyworth's document was a tall, gregarious Air Force Colonel by the name of Gilbert D. Rye. While George Keyworth had spent thirteen years as a research physicist, working his way up to the modest position of Division Leader at the Los Alamos National Laboratory, Colonel Rye had spent seventeen years learning the skills of trench warfare in the military bureaucracy.

Rye was a member of that new breed of military officer skilled in the politics of administration and the management of large operations. During his obligatory tour of duty in Southeast Asia, he had served as Deputy Director of Budget for the Air Force in Vietnam. After Vietnam, he had served as an assistant to a variety of commanders, at Kirkland Air Force Base in New Mexico, at the Space and Missile Systems Organization in Los Angeles, and finally at the Supreme Headquarters for the Allied Powers in Europe. Rye had climbed into the Air Force officer corps the hard way: not through the Air Force Academy, not through the officer training program, but as an enlisted man. He began his military career as a technician on a flight line, where his performance earned him an Air Force scholarship to the University of Colorado and a commission upon graduation. Learning as he went, Rye worked his way up to the position of Chief of the Space Plans Division in the headquarters of the United States Air Force in Washington, D.C.[11]

In his position as Chief of the Space Plans Division, Rye had received an invitation to attend an obscure conference, near New Orleans, being held to discuss the possibility of initiating plans for an outpost in space. He had accepted the invitation, the only person to attend NASA's November 1981 Space Station Planning Workshop who was not an employee of the civilian space agency. The space station idea fascinated Rye, even though the Defense Department had not taken a position in favor of the project. At the conference, Rye listed twelve ways in which the Defense Department could use the space station, suggested five ways in which NASA could enhance the survivability of a space station in the event of war, and called the concept of a modular space station "appealing."[12]

Five months later Air Force officials detailed Rye to the White House. He would, they told him, represent the interests of the defense community as Director of Space Programs on the staff of the National Security Council.

Among all the councils and interagency groups populating the White House galaxy, the National Security Council gave off the most light. In essence, the National Security Council served as the President's War Cabinet. Established in 1947 in conjunction with the unification of the Department of Defense, the National Security Council gave the President a cabinet to which he did not have to invite irrelevant department heads when the United States had to act in its more-or-less permanent state of war. Establishment of the council led to the establishment of a staff, which gave the President a source of information independent of that fed to him by civil servants in the permanent departments of peace and war. The staff, officially listed at fifty-seven employees plus detailees, worked under the Assistant for National Security Affairs, one of the President's closest advisers.

The Secretaries of Defense and State, statutory members of the National Security Council, liked to stack the council staff with detailees expected to promote the departmental point of view. Once ensconced on the White House grounds, however, the opportunity to order generals, admirals, and ambassadors around tugged on the loyalties of the detailees. From his office on the third floor of the old Executive Office Building, Gil Rye looked down on the West Wing of the White House and the Oval Office, a marvelously powerful view. Rye had access to the Oval Office through his new boss, William P. ("Judge") Clark, the President's Assistant for National Security Affairs, and a career of cultivating the needs of the people for whom he worked.

Rye arrived at his desk and ordered George Keyworth's assistant in the White House Office of Science and Technology Policy to send over a copy of the new space policy for review. In it came. Rye requested the copy on the grounds that the study had been directed by the Assistant for National Security Affairs who preceded Clark. Such a directive might have existed, but Rye did not have it. He was bluffing. "Had OSTP insisted on justification for sending the space policy document through the NSC system," Rye explained, "I would have been hard put to provide it." Keyworth should have sent the proposed document directly to the President, Rye said, "as he had every right to do." Rye took the draft of the new space policy to his boss, William Clark. Rye argued that Clark should assume the chairmanship of the Senior Interagency Group for Space. "I wanted to ensure that we controlled the space agenda and not the science adviser," Rye said. "Neither Jay [George Keyworth] nor Vic Reis [Keyworth's assistant] were strong space advocates and certainly did not have the access to the President necessary to push through many of the critical decisions lying ahead, especially space station."[13]

William Clark did. Clark had started to work for Ronald Reagan when the President was a private citizen campaigning for the governorship of California. Once elected to the statehouse, Governor Reagan appointed Clark as his chief of staff, then gave him a series of judicial appointments culminating in Clark's appointment to the California State Supreme Court. When Reagan moved to Washington, Judge Clark did too, along with other members of the California crew. An old and trusted friend, Clark had easy access to the President, with whom he met daily to discuss political and national security affairs.

The night before the President's July 4, 1982, speech at Edwards Air Force Base, Judge Clark had carried the new space policy to the President at his ranch in the hills above Santa Barbara. The document named the national security chief as chairman of the Senior Interagency Group for Space. Gil Rye waited in a motel at the bottom of the hill in case anything went wrong. Neither Clark nor Rye had been present when George Keyworth started to draft the new space policy. Clark was helping run the State Department, Rye was working in the Pentagon. But when the space policy, with its newly revised coordinating group, went to the President for his signature, Clark and Rye were on the scene and Keyworth was not.[14]

Reagan had approved the document. In getting Reagan's approval, Clark and Rye deprived Keyworth not only of the chairmanship of the group but also of his status as a voting member. Then they cut out David Stockman, taking away his vote as well. "Representatives of the Office of Management and Budget and the Office of Science and Technology Policy will be included as observers," the presidential directive read.

Within the eight-person Senior Interagency Group for Space, the national security community held five votes. The Department of Defense, the Joint Chiefs of Staff, the Central Intelligence Agency, and the Arms Control and Disarmament Agency sent voting members to join with the President's chief assistant for national security affairs. National security interests dominated space policy under the new coordinating group. For programs like the space shuttle, behind which the civilian leaders of the Defense Department had thrown their weight, the new interagency group provided NASA executives with a much friendlier route to the Oval Office than forums like the White House budget review process.

To get a favorable recommendation for the space station up to the President through SIG (Space), Jim Beggs would need the support of the Department of Defense. Besides himself, the only other member of the coordinating group from a civilian agency was the Deputy Secretary of Commerce. The eighth member, the representative from the Department of State, was unpredictable.

With the support of the defense community, a recommendation endorsing the space station could go from the Senior Interagency Group for Space to the President, just as if it had come out of the cabinet. It would not have to go through the Office of Management and Budget; it would not have to go through the White House Office of Science and Technology Policy; it would not even have to go through the cabinet. It might have to go through the National Security Council, but that would prove pro forma, since the national security community dominated that too.[15]

The directive that Reagan issued on April 11, 1983, ordered the members of SIG (Space) to complete a study of the space station "not later than September 1983 prior to presentation to the President," just five months away. In a two-page memorandum, drafted by the White House staff, Reagan told his interagency council to study four options. As alternatives to NASA's desire for a "space shuttle and a fully functional space station," the group would consider the pros and cons of "an evolutionary/incrementally developed space station," tended by astronauts part of the time and unoccupied the rest; "unmanned platforms" delivered into orbit by the space shuttle; and the current combination of shuttle flights and "unmanned satellites."[16] Ronald Reagan personally signed the directive. By this he meant to impress the members of the interagency group that their study would provide the basis for a presidential decision. He was polling them for their advice, but he wanted the study to come back to him so that he could make the final decision.[17]

The people who sat on the Senior Interagency Group for Space, such as NASA Administrator Jim Beggs, would not draft the study report themselves. The presidential directive established a working group to do that. "The Working Group will produce a summary paper that assesses the issues and identifies policy options," the directive read. The working group would draft the report; SIG (Space) would discuss it, vote on it, and send it to the President for a final decision.[18]

The presidential directive allowed each of the voting members of SIG (Space) to appoint a representative to the working group. It then gave NASA the grand prize. The President asked the NASA Administrator to appoint the chairman of the working group. Beggs appointed John Hodge, the director of NASA's Space Station Task Force, and told Daniel Herman to go along as Hodge's deputy. Not only would John Hodge get to make his pitch to the working group, he would have overall responsibility for drafting the report that would go to President Reagan for a final decision. After more than twenty years of trying to push a space station decision into the Oval Office, NASA had finally found a way to get its plans onto the President's desk.

Beggs worried, however, that he might not be able to win Defense Department support for NASA's "fully functional space station." If the national security community threw its weight behind another option, Beggs would come under enormous pressure to accept a compromise so that the group could take a recommendation to President Reagan on which they all agreed. Beggs did not want to negotiate away NASA's vision of permanent occupancy. Anticipating that he might lose the defense community's support, Beggs remembered an anecdote from the early days of the aerospace industry, told about J. H. "Dutch" Kindelberger. "Kindelberger ran North American Aviation for many years," Beggs recalled. "He was a legend in his own time." Kindelberger designed famous airplanes "that the Air Force thought they did not want. Dutch would build them and get them designed and then the Air Force always found that they needed them."[19]

"When he'd see a young engineer," Beggs said, Kindelberger would draw the engineer aside and say, "Now son, what you want to do is make sure that you've got a way out of every problem. Don't just design down a single track. If you run into trouble, you've got to have a way out."

"And he would ordinarily look at him and say, 'Let me illustrate that. Do you know how to pull a rabbit out of a hat?'

"And inevitably the engineer would way, 'No sir, I don't. How do you pull a rabbit out of a hat?'

"And Dutch would say, 'You put a rabbit in the hat.'"

"Young engineers," Beggs continued, "tend to get intent on doing it exactly the right way and they forget that sometimes you run into a blank wall."

"I kept telling the study group, give me enough options, that when you run into things like the DOD not wanting any part of it we can shift our feet very quickly and move off in another direction. Don't get tied down to a single idea or a single approach or a single configuration."

If NASA could somehow put that rabbit in the hat, Beggs thought, the President would surely pull it out.

16

The Rabbit in the Hat

"Can I have the first viewgraph?" John Hodge asked as the lights went down. As director of NASA's Space Station Task Force and now head of the SIG (Space) working group, Hodge repeated the space station story frequently during the spring and summer of 1983. This presentation, made before the members of the House Subcommittee on Space Science and Applications, was typical of the pitches he made. He made similar presentations to NASA executives, to the White House staff, to other lawmakers, to supporters of the space program, and to the members of the presidential working group that he chaired.[1]

"The beginning of the space station task force and the intensive effort that we started on planning began after the first four shuttle flights were completed." [Four photographs of the spaceship Columbia, landing after each of its four test flights in space, appeared on the screen.]

"As you know, Mr. Beggs had stated in his confirmation hearing that he felt that the next logical step was the space station," Hodge said. "In addition to that, Mr. Reagan, our President, last year on July the 4 issued a national space policy." [The pictures of the space shuttle disappeared and the President's words committing the United States to "permanent space facilities" took their place.]

One year had passed since the President had announced his desire for "a more permanent presence in space." The Space Station Task Force had been filling in the details ever since, with Hodge and his compatriots describing the details through the "top twenty" viewgraph presentation.[2]

"It is important for us to recognize that when we made the decision in the early 1970s to go ahead with a shuttle program, then, in fact, at that time, we thought in terms of a routine access to space. And that was a combination of both the shuttle, the transportation system, and the space station itself."

"That same decision had to be made in the Soviet Union. And it is also apparent that they decided, probably because of a lack of technology, that they would go ahead with the space station rather than the shuttle." [Pictures of the Soviet Salyut space station flashed on the screen.]

"What that resulted in," Hodge explained, "is us having a very good capability to provide routine access to lower Earth orbit through the shuttle system, which has been extremely successful. Whereas, what they have done is begun to understand man's role in space with a very, very extensive set of capabilities over the last ten years."

"I think if you ask the public at large, and quite possibly most of the people within NASA what a space station was, they would think in terms of the movie that came out 15 or 20 years ago." [A viewgraph revealed a modest facility, assembled around a small module, not a large rotating wheel as depicted in the movie *2001: A Space Odyssey*.] "That was a very large rotating vehicle with 100 people on it," Hodge said, referring to the space station in Stanley Kubrick's film.

"We believe that a space station is, in fact, many things." [He showed a viewgraph splattered with words, not pictures.] "There has been a tendency for the discussions and the debate over the last year to concentrate on whether we should have a man, or should he be replaced by unmanned systems." [Hodge flipped up a picture of the space shuttle with its cargo bay open to the heavens, the earth floating by below.] "We believe that that is the incorrect debate," Hodge observed, pressing forward with the key element in the plans he and the members of the Space Station Task Force had developed:

"Conceptually, we see the space station as both a manned element and unmanned elements, co-orbiting so that they have access to each other. The manned base would have the things that you would expect—somewhere to live, somewhere to work, somewhere to do things, somewhere to operate the station itself—and an ability to expand as the capability requirements expand during a decade. In addition to that, we see the need for large unmanned platforms, either single-purpose platforms, such as a space telescope, which is already under construction, or platforms which contain multiple instruments." [A picture of the whole earth appeared, with lines delineating different orbits encircling the globe.]

"These things do not actually fall into a single inclination," Hodge explained. "What you then need is the ability to go backward and forward between the manned and unmanned elements. That is sort of a mini-transportation system." In future years, Hodge continued, the space station would require "a large high-energy upper stage which will allow you to go from that transportation base to the higher-energy orbit such as geosynchronous or back to the Moon or to Mars, whatever it is you want to do."

"For the past year, we have been concentrating on the why of the space station rather than the what," Hodge said as he laid out NASA's rationale for the facility. "It is obvious that unless we build a multifunctional facility, it really is not going to be justifiable."[3]

The preference of the members of the Space Station Task Force for a multi-functional facility combining an occupied core with automated platforms had been re-enforced by the work of the aerospace contractors preparing the mission analysis studies. In April 1983, as President Reagan issued his directive setting up the White House study group, Hodge received the contractor studies. Eighty-one volumes in all, the reports made a huge pile on the floor. "This is the why," said Hodge, referring to the missions the space station could perform.[4]

The contractors had used a variety of techniques to identify the interests of what Hodge called the user community. The Boeing Aerospace Company gave a subcontract to RCA Astronautics to investigate the uses of a space station in launching communications satellites manufactured by RCA. The Lockheed Corporation held a series of workshops to which it invited a broad segment of the user community, explaining the capabilities of the space station and inviting the attendees to respond.[5]

In the spring of 1983, Hodge gathered up the mission analysis studies, along with the chief people from the Task Force and its working groups, and took them to NASA's Langley Research Center near the mouth of the Chesapeake Bay for an eight-day retreat. "We had 42 people down there two weeks in May, from the 6th to the 13th. They included industry." Representatives from the Air Force also attended, since they were being asked to identify military missions that could be performed on the space station, as did international participants, who were conducting their own mission analysis studies.

"They produced a set of realistic missions that are likely to take place during the 1990's," Hodge said of the various working groups assisting the Task Force. "They concentrated on science and applications, commercial, and technology development missions."[6] Brian Pritchard, the Task Force member responsible for the mission analysis studies, loaded 107 missions onto a computer program at the Langley Center along with the requirements for performing them.[7]

Scientists wanted to send a robot probe to the surface of Mars before the end of the twentieth century. The robot would roam about, collect samples, and return. The robot and its spacecraft could be assembled in orbit; on its return, the samples could be analyzed in the quarantined environment of the space station to guard against Earth-based contamination. Pritchard insisted that the mission could be performed only from a space station, although the Task Force held to the less aggressive position that the space station would simply "facilitate" the mission to Mars. Pritchard loaded it onto his computer as "SAA 0110" (the "SAA" standing for "Science and Applications").[8]

Scientists also wanted to build a large deployable reflector above the atmosphere of the earth, from which they could study infrared signals coming in from the depths of space. The best reflectors on the surface of the earth were big, as large as the 1,000-foot-wide Arecibo Ionospheric Observatory in Puerto Rico. To deploy a reflector wider than 10 meters in space, astronauts would have to assemble the dish in the vacuum surrounding the space station and push the reflector out to a free-flying orbit. Pritchard named that mission "SAA 0020" and loaded it onto the Langley computer.[9]

Aerospace contractors thought that the space station would excite manufacturers to enter space, especially with NASA footing part of the bill. Without a space station, entrepreneurs could submit proposals to make pharmaceutical products or grow new crystals in space, using the space shuttle, but it took ten years just to make a proposal, schedule an experiment, fly it, and analyze the results. Members of the Space Station Task Force argued that a continuously operating materials-processing laboratory would speed up that process significantly. "A manned space facility operated on a continuous basis," the McDonnell Douglas Astronautics team wrote, "is essential if we are going to develop the emerging commercial opportunities in space." The "MacDac" team predicted a potential world-wide market of 250 million people for twelve space-manufactured pharmaceutical products. Pritchard's team was more cautious, since everyone was working from so few "data points." Nonetheless, it scheduled twelve materials-processing missions on the space station, from pharmaceuticals to fiber optics, and loaded them onto the Langley computer.[10]

Nearly one-third of the missions planned for the space station consisted of what the Task Force called "technology development missions." Technology-development mission 2060 required NASA to develop techniques by which astronauts could construct large objects in space. Technology-development mission 2070 required the space agency to perfect "lightweight, flexible space structures that cannot be tested on the ground." Technology-development mission 2120 challenged NASA to experiment with laser technology, which in turn might open new frontiers for the generation of electric power or space-based propulsion.[11]

The 107 missions that Brian Pritchard loaded onto the Langley computer required a large space station, larger than John Hodge could ask the White House study group to bless. NASA sought approval for what in 1983 it called "initial operational capability," a facility from which as few as 33 missions could be performed during the first three years of operation. Other missions would be completed later, during the next seven years, as the space station grew. NASA's ability to present an incrementally "phased in" space station helped win support for the facility. Delicately, NASA officials worked to justify a space station that ultimately could perform 107 missions but that initially would cost only enough to perform a fraction of that number.[12]

Without exception, none of the 107 missions on Brian Pritchard's computer were new. They were missions already planned for space. The aerospace contractors and Brian Pritchard's working group went to people who were interested in space science, commerce, and technology development and asked them to list the missions they were planning anyway. Pritchard, Hodge, and the other members of the Task Force then identified those missions that would be enhanced by the existence of a space station.[13]

With the possible exception of the Mars Sample Return, none of the 107 missions absolutely required a space station to be performed. If necessary, they could be carried out on unmanned platforms or from the space shuttle, especially if the shuttle could be outfitted to remain in orbit for longer periods of time. A large number of missions—fully one-third of the missions planned for the initial facility—would not even be performed from what most people thought of as the space station. They would be conducted on orbital platforms or from free-flying facilities physically separated from the main space station. Some, such as the Hubble Space Telescope (SAA 0012) could be serviced from the space station but would be operated from the ground. Others, such as the earth science research program, would be mounted on a fully automated platform flying over the North and South poles. The polar platform could be reached from the main space station only by returning to the surface of the earth. As John Hodge had said, the Task Force could not justify a single-purpose facility, circling the earth like some sort of airport in space.[14]

To fully justify the space station and push the number of missions past the critical point, NASA officials worked to include military missions on their list. The Department of Defense spent more money each year on space operations than NASA did. The Department of Defense had booked one-third of the flights of the space shuttle to deliver satellites into orbit and perform research in space. The national security community had more satellites than NASA did—satellites for tracking troop movements, satellites for detecting rocket launches, satellites for verifying compliance with arms treaties, satellites that captains of nuclear submarines could use to aim their missiles, navigation satellites, early-warning satellites, and communication satellites.[15]

Potential national security uses for a space station fell into a broad range of categories. Branches of the military could use the space station to store and launch satellites. They could use it as an observation platform. They could use it as a military command post in the sky. They could use it as a switching station from which to collect and verify data from unmanned satellites. They could use it for a wide range of research-and-development activities, everything from infrared surveillance technologies to experiments with space-based weapons that might be able to shoot down inter-continental ballistic missiles.[16]

The Air Force set up three panels to study possible military uses of a space station. They contributed $300,000 to Brian Pritchard's working group, money that Pritchard turned over to the eight aerospace contractors to study national security needs. Pritchard's people met with members of the Air Force Systems Command Space Division, the aerospace contractors met with them, and the contractors went out to different military commands to discuss defense needs. "The final conclusion," Pritchard said, "was that they saw no specific Department of Defense need for a space station at that time."[17]

"There are no currently identifiable mission requirements which could be uniquely satisfied by a manned space station," Air Force officials announced. "Experience gained in NASA's manned space program to date," the study team wrote, "has not altered a view that most military space missions may be better performed by unmanned systems." Technological developments in high-speed computers, video links, and satellite relays, they noted, "have vastly increased our capability" to operate unmanned systems from the ground. "No requirements were found where a manned space station would provide a significant improvement over alternative methods of performing the given task."[18]

"All of their activity in space was on unmanned vehicles and they concluded that they had no unique requirement for a space station. I think that is significant," Brian Pritchard stated, "the word 'unique.'" The Air Force looked at missions that could be performed only from a space station, uniquely, such as a command post in the sky, which was obviously too vulnerable to enemy attack. "They got hit with a partial bill on shuttle," Pritchard suggested, having failed in his attempt to get the Air Force to contribute missions to the list. "They would much prefer that NASA spend all its money to develop a facility such as space station and then if they suddenly should find a need to use it, well and good."[19]

Pritchard submitted his list of missions in June 1983, without any Defense Department activities on board. "Potential National Security missions are omitted," Pritchard's report read, "but are expected to be included in later drafts." NASA executives continued to lobby top Defense Department officials, pressing them to alter the Air Force position. Jim Beggs and Hans Mark still believed that they needed the support of the Defense Department to smooth the path of the space station through the White House. They were not prepared to give up their search for that support just because the Air Force refused to submit a list of missions to Brian Pritchard's working group.[20]

Hodge continued to build the rationale for the space station on missions, even though the Air Force continued to withhold theirs from the NASA plan. He shied away from playing what the leaders of the Task Force called the "communist card," the argument that the United States needed a space station because the Soviets had one. America's last space race with the Soviets had left NASA without a permanent base in space or a rocket large enough to launch it. Hodge avoided arguments about the cost effectiveness of a space station, even though each of the eight mission analysis studies contained a section on that subject.[21] During the approval of the Space Transportation System, debates over cost had turned into compromises over design. In any case, if Hodge brought up the cost advantages of the space station, the opponents would use this as an opportunity to point out that NASA had still not met its cost goals on the shuttle.

As Hodge continued to stick with missions as the basis for his space station plan, officials from the U.S. General Accounting Office notified NASA of their intent to issue a special report challenging the mission approach. Technically an arm of Congress, the General Accounting Office had been created in 1921 as part of the same legislation that set up the President's budget office. For decades the GAO, as it was known, drifted along content with the job of auditing the financial practices of federal agencies to make sure that the bureaucrats spent funds in the manner intended by Congress. In the late 1960s, however, under the leadership of Comptroller General Elmer Staats, the GAO began to conduct a large number of what became known as "performance audits." The analysts at GAO delved into the management practices of big government bureaus and published fat, devastating reports on government fiascoes and mismanagement.

No one on the Space Station Task Force seemed to know who had requested the audit on the space station proposal. It could have been a hostile congressman or an unhappy aide on a congressional committee. Still, the auditors were there, standing at the door, anxious to come in and ask questions about NASA's space station plans.

The thrust of GAO's challenge to NASA's mission model appeared on page four of the first draft of the special GAO report. "Planning for a major new development program," the auditors wrote, "should include an indepth analysis of the relative costs, benefits and risks of alternative projects." The auditors offered examples of possible alternatives: a more aggressive planetary program, an expanded space applications program, a concentrated effort to increase the efficiency of the shuttle, or the development of a vehicle that could take satellites from the shuttle's low-earth orbit out to geosynchronous orbit at 22,000 miles.[22]

"NASA officials responded to our inquiries about alternatives," the auditors complained, "by dwelling on the merits of a space station." In the second draft of their report, issued in April 1983, the auditors repeated the charge that "NASA is not considering alternative new starts in focusing its planning on a space station as its next major development." That was the weakness in the mission model; it assumed the need for a space station, sending the contractors and the members of Pritchard's working group out in search of missions to be performed on it.[23]

In their reply to the draft audit, Task Force members defended the mission approach to space station planning. Their response deftly sidestepped any discussion of alternatives. NASA had known that it needed a space station since "the end of the Apollo program." The space shuttle, moreover, "was conceived as a transportation system for a space station." By focusing on the missions to be performed on the station, Task Force members promised the auditors, "the analysis will not be skewed in favor of a station."[24]

The GAO audit echoed many of the claims that opponents of the space station were using against the initiative. NASA was not considering alternatives. The space station would cost too much. It could crowd out other programs, principally space science. Most GAO audits, however, did more than repeat claims. They provided ammunition, in the form of hard data and tight analysis. And they required the head of the affected agency to respond formally to the audit in front of the congressional committees that oversaw the program.

This audit did not. Written in the form of a letter rather than a full-blown performance report, the document did not back up its claims with details. In a style untypical of GAO audits, it tended to be polemic, apparently a result of poor staff work. When the polemics got edited out of the second draft, so little substance remained that the letter was never sent. As a result, NASA never had to reply formally to the GAO charges.

With the GAO audit out of the way, Hodge could do what he had always intended to do with the mission analysis studies. He let his engineers pick up their pencils. For nearly one year Hodge had maintained his ban on drawing pictures. Even doodles, said one participant, "were destroyed at the end of the day to make sure that nothing got out that might be suggested as the NASA-approved design for a space station."[25] During the first year, the two working groups that Bob Freitag had established to study space station architecture and requirements had found little work to do. The groups had examined issues like alternative shapes for reducing drag, but had not delved into much detail.[26]

Now waist-deep in his efforts to explain the space station to the White House study team, John Hodge needed architectural details and he needed them fast. Without some preliminary architectural details, he could not calculate the cost of the space station. Without knowing the cost of the space station, he could not tell the members of the White House staff and the SIG (Space) working group what sort of a program the money would buy. Hodge did not need blueprints, but he did need a more precise definition of the elements that would make up the facility.

By April 1983, when the SIG (Space) working group began its analysis, the Space Station Task Force had not grown much past a twenty-five-member team, still packed onto the fifth floor of the main NASA headquarters building. Hodge knew that he needed another thirty-five to forty people, working full-time at one location, to pull together the preliminary architectural details and cost-out the space station. But he had no place to put them even if he could get them.

Hodge and Frank Hoban finally found some space, two blocks away on the fourth floor of the General Services Administration Building, with no windows and a leak in the ceiling that required a worker to remove a 55-gallon drum of wastewater each day. "Just like a bunker," said Luther Powell, the NASA engineer Hodge shanghaied to lead the effort. "I didn't want to go to Washington," Powell said. He had spent all of his twenty-three-year NASA career working at the Marshall Space Flight Center and nearly all of his life in Alabama. Born in Bessemer, Alabama, one of four children of a working-class electrician with a third-grade education, Powell had dropped out of high school himself to join the Navy.

The Navy had sent Powell, a mischievous-looking boy with an attraction to trouble, to electronics school. "I was kind of a rebel," Powell said. He had a flying sergeant from the Marines as his proctor. The sergeant examined Powell's work and told him that "if you get your head out of your rear you might do something if you just put your mind to it." Powell completed his tour with the Navy, went back to Alabama, finished high school, and worked his way through college with a wife and two children. He came to Huntsville during the von Braun era in 1956 to work for a contractor on the Centaur rocket and rarely left town.

"I think the first idea I had of the situation was from John Hodge," Powell remembered. "He was looking for a way to do some systems engineering of the different working groups."

"Quite naturally, I declined." The people at the field centers were still not convinced that the Space Station Task Force would be able to get the program approved. "It wasn't too much longer after that that Phil Culbertson asked me about coming. And then the next thing happened, my center director told me he had discussed it and agreed with Phil that I would go."

Powell arrived in Washington on April 24, 1983, together with the team of three dozen engineers he had pulled together from the different NASA field centers, all committed to spending one year in Washington working on the space station initiative. Hodge told them to take the materials produced by Freitag's working groups and the mission analysis studies and "define what a space station ought to look like, how much it ought to be able to accomplish given the requirements we were able to derive so far, and how much it would cost." Powell put up a sign outside the door of their dismal quarters: "Space Station Concept Development Group."[27]

Hodge and Beggs had been going around town talking about a space station that would cost $4–$6 billion at initial operational capability, a space station that could sustain a permanent human presence in space and conduct the first thirty-three missions identified by Brian Pritchard's mission analysis working group. That was clearly more than the $3-billion space station the Fletcher committee had proposed and considerably more than the less-than-$2-billion space platform Powell's own center had been pushing.[28] Powell looked at the initial set of missions, the technology, the requirement that the space station contain both a live-in core and unmanned platforms, and concluded that the space station as the Task Force leaders then conceived it would cost substantially more than $4–$6 billion.[29]

The electric power requirements surprised Hodge. Powell had done what Hodge had asked him, taken the missions that Brian Pritchard had laid out, calculated their power requirements, and distributed the missions so as to reduce total electrical load. Even by conducting operations in shifts, Powell could not get the initial power requirements down below 55 kilowatts. Skylab, by contrast, had limped along with 7.5 kilowatts of continuously usable electric power. "This was quite in excess of what we had expected," Hodge admitted. "Needless to say, the first time you go through that, the requirements for the thing are far in excess of what the potential fiscal capability is going to be."[30]

Powell's engineers designed the power system in 12.5-kilowatt packages. They started with four. "That didn't live very long," Powell observed. Materials processing, science, and space station servicing required more. Within a few months, they were up to 60 kilowatts, then to 75. The space station, they calculated, would start at 75 kilowatts and grow to 160 kilowatts by the year 2000.[31]

Working from Brian Pritchard's mission list, Powell and Hodge began to calculate the size and shape of the modules around which the space station would be constructed. Once again, the list of missions advanced the state of the art. Skylab had been launched into space as one large cylindrical can. The space station, just to meet the initial set of missions, would have to be constructed in space out of many different modules.

The crew, six to eight men and women, would live in a habitat module. They would work in a laboratory module. Supplies would be stored in a logistics module. Task Force workers planned to house the power and thermal equipment in a utility or resources module. To link the modules together, they examined the concept of what Powell called a "multiple berthing adapter," a module that served as a central hallway. For the initial set of missions, Hodge and Powell calculated that they would need at least one of each module.[32]

Jim Beggs kept telling Hodge and Powell that a space station was something one bought incrementally, "by the yard." Beggs wanted to start with a basic facility that would allow people to stay permanently on board plus a list of options that he could add, depending on the amount of money he could get the President to approve.

The first time Powell met with Beggs, Powell said, "I'm going to show you the bolt of cloth from which you will buy a space station by the yard." Powell suggested that different modules be constructed out of one common frame. "Three of us sat together one afternoon and came up with that," Powell recalled. "We all thought we were crazy," he said, knowing that NASA's field centers would insist that each module needed to be different. The concept of the common module, however, would reduce costs and make growth easier. Powell thought "that there wasn't anybody in the agency that had enough clout to make it stick."[33]

In order to create a multi-functional facility, Powell examined the requirements and estimated the costs of two unmanned platforms, defining them as part of the overall space station. One would fly in the same 28.5° orbit as the main station, and the other would travel in a 90° orbit around the North and South poles. Powell's Concept Development Group developed plans for a remotely controlled pod that astronauts working inside the main station could use like a robot to service nearby satellites and platforms. The group added a few mounts on one end of the space station where four large experiments could be installed. They also made plans for an attachment point on the main space station where satellites could be held for repair.[34]

These elements made up the space station as the leaders of the Task Force envisioned it during the summer and fall of 1983. The total cost, Powell figured, would come out somewhere between $7 billion and $12 billion, depending upon the options and the sophistication of the system. Hodge even allowed the engineers to draw pictures of it, primitive by comparison to the detailed configurations yet to emerge, but pictures nonetheless.[35]

"Where are we now?" asked Hodge, turning to the last of the viewgraphs. "We have all of the centers participating. We have a steering committee which is helping us with the technology. We have the user communities involved. We have taken a first cut at what this all means.

And we are beginning to understand the necessities of the management process."

"When we're through, we want a clearly defined concept, we want a management process that we know will work, we want a detailed design, and we want to have industry on board and working with us. The reason we want to take that deliberate approach to this is so that the systems are well understood, that we know the technology will work, and we have a good understanding of how much it is going to cost."

"And what that will allow," Hodge concluded, "is for the executive and the legislative organizations to make an informed decision about whether we should proceed with the program."[36]

17
SIG (Space)

John Hodge looked at the people seated around the table in what he called NASA's wire room, also known as the cage. Gil Rye had told Hodge and the members of the SIG (Space) working group to produce a report by the first week of July 1983, and Hodge had assembled the group to finish the job and vote on the summary report, line by line. Representatives from the national security community insisted upon meeting in a secure room, one designed to thwart any electronic surveillance. Hodge reserved special chambers on the seventh floor near the NASA Administrator's office to assemble the study for the Senior Interagency Group for Space.

The national security community held the balance of power in the working group. The Department of Defense, the State Department, the Central Intelligence Agency, and the Arms Control and Disarmament Agency held four of the group's six official seats. Hodge had also invited representatives from the White House science office and the budget office, on the assumption that it was better to have them criticizing the proposal from inside the group rather than standing outside and fighting it. If he could win the support of the national security representatives, adding their votes to his own and that of the representative from the Department of Commerce, Hodge could send a favorable recommendation to the President through the Senior Interagency Group for Space, even though the scientists and budget analysts would dissent.

The prospects seemed dim. The national security community, Hodge observed, had stacked the working group with people from the Air Force. The Defense Department had sent an Air Force officer. One of the two representatives from the CIA was from the Air Force and so was the representative from the Department of State. Even the President's science adviser had sent an Air Force officer to the mid-level working group. Within the Air Force, most of the top-ranking officers opposed any military participation on a NASA space station.[1]

Meeting with his equals in an interagency group, Hodge faced a problem common to career civil servants trying to shape public policy from the bottom up. The other careerists in the group would support

NASA's goals only if the agency lowered them. Twelve years earlier, outsiders had force NASA to accept a significantly less ambitious space shuttle than the agency had proposed. Now they wanted NASA to accept significantly less than a permanently occupied space station.

Hodge was prepared to negotiate the scale of the space station, designing the facility, in the words of Daniel Herman, the other NASA representative in the working group, "to a cost ceiling." Hodge was willing to phase in future elements of the station and he was prepared to delay the bulk of the spending for two to three years until NASA had fully defined the requirements of the facility. He was not prepared, however, to negotiate NASA's primary goal: a space station with people permanently on board.[2]

The job of the other representatives on the working group, as Hodge and Herman saw it, was to move NASA off of this position—to discredit the "fully functional space station" and to force NASA officials into a backup position where they would have to accept something less. Something less might be a "man-tended" space station or simply the use of the space shuttle to do some of the work that a space station would do.

The science advisers continued to insist that the space station program, even with its unmanned platforms, would not significantly enhance the quality of the science on the missions that Hodge and Pritchard had laid out. Those missions, they argued before members of Congress and NASA's potential allies, could just as easily be run from the space shuttle and probably at less cost.[3] The Space Science Board of the prestigious National Academy of Sciences made the argument official. "The Board," its members wrote to James Beggs, "examined the set of specific missions proposed for implementation from the space station system during the years 1991–2000. It has found that few of these missions would acquire significant scientific or technical enhancement by virtue of being implemented from this space station. In view of this and the adequacy of the present space transportation system for the purposes of space science, the Board sees no scientific need for this space station during the next twenty years."[4]

Victor Reis, Assistant Director of the President's Office of Science and Technology Policy, set a neatly constructed trap into which NASA officials could step. "With one exception, each and every one of the missions can also be done—and done well—by a judicious combination of shuttle orbiters and satellites, if we extend the mission duration of the orbiters." As Reis pointed out, "A program of orbiters and satellites would cost less, perhaps considerably less, than the equivalent space station program. I believe this is so because of two reasons. First, the requirements for the wide variety of space missions are often so different that the potential savings from sharing platforms will often be more than wiped out by the additional cost of making them work on the same platforms. And second, sharing space with human beings is always costly."

"What it boils down to is this," Reis argued. "If you can do basically the same space station job with satellites and shuttle orbiters at a lower cost, are there any arguments for going ahead with the space station? There is one. And if it is neither technical nor economic, it is a powerful argument nonetheless." NASA was created in large part, Reis explained, "to do the macroengineering project Apollo, the manned mission to the Moon. Apollo was not a scientific experiment and it was not undertaken for economic return. It was a statement of national will in an era of intense international political competition. And it was successful, historically successful. NASA remains much the same organization today as in the days of Apollo. It is organized and staffed to carry out large, complex manned space macroengineering projects. The space station is just such a project and there is no other that fits the NASA mold quite so well. In short, the space station project will provide NASA the central focus and scale that it is organized to do."

"Rejection of the space station at this time," Reis concluded, "would certainly lead to a different NASA, one much removed from the legacy of Apollo." If NASA were to put the space station decision in these terms, he would be inclined to support it. "To my way of thinking," Reis said, "this is the fundamental issue of the space station decision. Is the additional cost of the space station worth maintaining NASA as a politically potent organization, one which might provide the nation with the potential for significant political and social benefit?"[5]

The presidential directive setting up the working group gave Hodge the latitude to justify the space station on such a rationale. The directive instructed Hodge to assess the "social impact of a manned Space Station" and to take into account its contribution "to the maintenance of U.S. space leadership," an invitation to argue the importance of maintaining a strong civilian space agency. Hodge could use that language to produce a study that both NASA and the White House science adviser might sign. To do so, however, Hodge had to admit publicly that the space station could not be justified on scientific grounds alone.[6]

If Hodge refused, the science advisers and their allies could challenge him to produce the cost and engineering studies demonstrating how a manned space station would be more effective than other alternatives for conducting scientific experiments. NASA officials believed that a space station, in combination with satellites and unmanned platforms, was "the most efficient and effective alternative" for conducting space science and other missions in space.[7] Hodge had in his possession studies supporting that belief. Those studies, however, had been prepared by the aerospace companies that would build the multi-billion-dollar space station, which were not unbiased sources. If Hodge put those studies on the table, the scientists and budget analysts could demand an impartial study, delaying the initiative and pushing the cost issue up front.

It was a cleverly laid trap. If Hodge entered it, he had to repudiate the mission analysis studies upon which he had based the space station rationale. If he avoided it, he might be forced into the sort of cost/benefit debate that had bogged down the space shuttle decision twelve years earlier. Hodge refused to approach the snare. He chose not to compromise the rationale for the space station on that basis, and so the President's science advisers made NASA a second offer, one they felt the agency leaders could not refuse.[8]

Representatives from the science community thanked Hodge and the Space Station Task Force for involving them in the planning effort. "This is in contrast," they said, "with what happened when these communities were presented with the space shuttle and spacelab, two previous systems imposed on them."[9] While maintaining the argument that the space station would serve "no scientific need" through the year 2000, the scientists held out the possibility of a compromise. "In the longer term," the investigating group from the National Academy of Sciences wrote, "the Space Science Board sees the possibility that a suitably designed space station could serve as a very useful facility in support of future space science activities. Such a space station could provide means for erecting and fabricating large and novel structures in space, and for servicing, fueling, and retrieval of payloads in orbit. If NASA wishes to develop plans for such an ambitious and technically demanding space station for the next century," the scientists wrote, "the Space Science Board would be pleased to work with NASA in defining the properties of such a space station."[10]

As Hodge's study group finalized its report, George Keyworth traveled to Seattle to deliver a speech to the Joint Propulsion Conference, an annual gathering of aeronautical, astronautical, automotive, and mechanical engineers. "I'm going to throw you a curve this morning," said the President's chief science adviser on June 27, 1983. "I'm going to skip beyond specific topics in propulsion systems and jump to one of its grandest consequences—the U.S. space program." The government, offered Keyworth, was "near an important decision point in determining the future directions of our space efforts. In the past few months I've been giving a lot of thought to this subject," he said. "I'd like to share some ideas with you today—and I'd like to challenge the aerospace community to do some bold thinking about the future."

"How can we use our investments in space to bolster our science and technology base and to re-ignite the spirit of adventure that captured

America in the past?" he asked the participants. "The most-often discussed—and most immediate—possibility is a manned space station, serviced by the shuttle and composed of modules supporting a broad spectrum of space activities. This would obviously be a challenging use of the shuttle's capabilities, and the needed technology appears to be within reach." The persons backing the space station, however, "readily acknowledge that it is, in truth, only an intermediate step in a more ambitious long-range goal of exploring the nearby solar system. Why, then, can't we be forthright and lay those ideas on the table?" he asked. "Surely we learned from our Apollo experience that this country can be strongly supportive of ambitious new programs if the citizens share the vision and if they're convinced that important national needs are being served."

Touching the depths of NASA's long-range vision for the space station, Keyworth said that as part of such a compromise he could support the development of an orbital transfer vehicle to ferry astronauts from the space shuttle to high geostationary orbits, a permanent station on the moon, "or even manned exploration of Mars." He challenged the visionaries in NASA to ask for what they really wanted.[11]

NASA's leaders knew that they had in the White House a president more supportive of space activities than any chief executive in the previous twenty years. That spring, on March 23, 1983, just sixteen days before he had met with Jim Beggs to set up the space station study, President Reagan had instructed the Department of Defense to proceed with a multi-billion-dollar Strategic Defense Initiative. Labeled "Star Wars" by a skeptical press, Reagan's initiative challenged the Defense Department to develop the technology necessary to detect and destroy enemy missiles in flight. The head of NASA's shuttle program, Gen. James A. Abrahamson, left NASA and returned to the Pentagon to direct the program.[12]

By the most conservative of estimates, the Strategic Defense Initiative would cost \$30–\$40 billion during the next ten years. For roughly the same amount, the civilian space agency could take a large step toward a base on the moon or a mission to Mars and gain an orbital transfer station in the bargain. There was just one catch. The people with whom Hodge was dealing would support such a program only if NASA put it off until the next century. In the meantime, NASA could "continue to explore the requirements, operational concepts, and technology associated with permanent space facilities," a fancy way of saying that they could not build the orbital station now.[13]

In preparing their mission analysis studies, Brian Pritchard and John Hodge had been very careful not to include on their list any missions beyond the year 2000, such as NASA's long-held vision of an expedition to Mars. "Those kind of missions," Pritchard recognized, "would be very, very expensive." The members of his working group were "worried about our credibility if we built a case for space station which was dependent upon missions in the post-2000 period," Pritchard said. Given their view of the federal deficit, NASA officials decided to stick with the incremental approach and press for a low-budget space station. Still, no one in NASA seemed to know how much civilian space President Reagan might really buy.[14]

Having failed to draw NASA out of position, the opponents of the space station fell back on another tactic, one they had constructed six months earlier when Hodge's working group first met. If NASA refused to compromise, neither would the other representatives. Hodge could present NASA's position to the President in the working group paper and they would present theirs. Hodge could then present rebuttals to their points and they would present rebuttals to his.

The summary paper, of course, would have to satisfy the "terms of reference" contained in the President's directive establishing the working group. Hodge had discovered, when his group first met on October 21, 1982, that the White House staff would not take a directive to President Reagan officially setting up the working group until the group first agreed upon its terms of reference. It took Hodge nearly six months, from October 1982 until April 1983, to draft them. NASA wanted one set of instructions, the national security representatives wanted another. Hodge worked twice as long to negotiate the terms of reference than to produce the actual study. When the study emerged, it became painfully apparent why the terms of reference were so important. The terms of reference allowed the working group to present three positions on five issues on four options, a total of sixty permutations, plus a series of supporting papers.[15]

As soon as Gil Rye saw the report, he realized that it could not be used as a basis for a presidential decision. With each side presenting its own position, there was hardly a breath of consensus in the report. Presidential option papers needed to be crisp and concise, boiling areas of disagreement down to a few key points and explaining the consequences of choosing one option or another. By forcing through these particular terms of reference, the opponents of the space station turned NASA's reluctance to compromise into a summary paper that could not possibly go to the President. The process, it turned out, had been doomed from the start.[16]

Just to get the issue before the Senior Interagency Group for Space, NASA needed a tighter report. To get a tighter report, it needed a smaller group with a stronger head. Gil Rye studied the situation from his office overlooking the West Wing of the White House. His White House superiors, Judge Clark and President Reagan, seemed to favor the space station. His superiors in the Air Force still continued to insist that the space station would serve no military needs. Rye pondered his position. Officials in the Department of Defense had sent him on a tour of duty with the National Security Council to advance his career. His job, from their point of view, was to make sure that the Defense Department got to comment thoroughly on any White House decision that affected the national security community before the decision was made. That was why Rye had not supported a presidential commitment to a space station in the Edwards Air Force Base speech one year earlier. It would have bypassed the White House interagency review process. The Secretary of Defense would have protested loudly and heads would have rolled on the National Security Council staff for not stopping such a speech in time for the Defense Department to state its views.

From the Defense Department's point of view, Gil Rye had already gone too far on the space station. He had advised people from the NASA Administrator's office and the Space Station Task Force of the political strategies needed to win presidential approval. He had used his time with the President to set up the April 7 meeting between Reagan and Beggs, since Beggs did not have as much access to the President as Rye did through his White House boss, William Clark. Convinced that the United States needed an orbital base to maintain its leadership in space, Rye had worked with Judge Clark to promote the initiative.[17] And he had written the presidential directive setting up the study group in such a way that the opponents of the space station could not keep the issue off the President's desk. The directive said that "results of the study will be presented to the SIG(Space) not later than September 1983 *prior to presentation to the President*." The study results would go to President Reagan, along with a decision memorandum, but first Rye had to produce a study that he could take to the President.[18]

Rye decided to form a second study group and head it himself. He asked NASA to send him a new representative, someone who understood the White House policy process and could work full-time on the study. The Administrator's office chose Margaret Finarelli. A newcomer to NASA, she had come to Washington to work as a scientific intelligence analyst with the Central Intelligence Agency after completing her master's degree in chemistry in 1969. From the CIA she had moved to the Arms Control and Disarmament Agency, where she negotiated chemical weapons limits with the Soviet Union. She had followed that with a term of service in the White House Office of Science and Technology Policy. A colleague invited her to join NASA in 1981 when Beggs became Administrator, working out agreements for space exploration with foreign nations from the agency's international affairs division.[19]

Rye simplified the options in the presidential directive, which had obliged Hodge to compare a "fully functional" space station to an evolutionary space station, unmanned platforms, and unmanned satellites. Rye boiled the issue down to three options. The President could commit to a permanently manned space station now. He could commit to the evolutionary development of the space transportation system, using power-extension packages and other devices to extend its duration in orbit. This was the fall-back position to which opponents of the space station hoped to push NASA. Or the President could defer commitment pending further study. Options two and three, from NASA's point of view, were a nice way of saying that the space station was dead.

Rye insisted that his drafting team produce a short report, about twenty pages long. He asked the Defense Department, the Central Intelligence Agency, and the Office of Management and Budget to participate on the team. He told the group they had to produce their report by the first week of August and he scheduled a meeting of the Senior Interagency Group for Space for Wednesday, August 10, 1983, to take up the space station initiative.[20]

On the drafting team, only Rye and Finarelli favored option one, the fully functional space station. The representatives from the Defense Department and the Central Intelligence Agency continued to insist that they had no need for a space station and that the initiative would divert resources from more important programs, such as the modernization of the nation's satellites for arms control verification. Representatives from the Office of Management and Budget continued to argue that the space station was a waste of money and that NASA had grossly underestimated the full costs of the facility.[21]

Finarelli and the people in NASA who were backing her up sharpened their rationale. Rye steered them away from arguments he felt would not impress the President. They avoided details on the costs and benefits of a space station. They described the Soviet commitment to a permanently manned space station, but did not highlight this in a dramatic way.

Together, Rye and Finarelli built the case for a fully functional space station on two grounds: the effect such an undertaking would have on the leadership position of the United States in space and the missions the station would perform. "A space station is necessary to maintain real and perceived U.S. leadership in space," they argued. A favorable decision "would be a reaffirmation to the world of America's commitment to technological superiority and to space leadership." Foreign governments were prepared to participate in the venture, and a failure to move ahead at this time would "send the wrong signals abroad with respect to U.S. commitments to space leadership and technological superiority." Rye encouraged Beggs and Finarelli to lean on Reagan's theme that "American is back" and his optimism about the future.

Rye and Finarelli also emphasized the missions the space station would perform. "The space station would enable extensive commercial exploitation of space." This would include commercial production of critical materials not available on the earth, such as extremely pure pharmaceuticals, as well as the deployment of remote sensing instruments for studying commercial resources on the land and under the seas. The space station would allow NASA to understand the ability of humans to function for long periods in space and thus lay the groundwork for "future manned exploration and exploitation" of space in the twenty-first century. The space station would also provide a permanent base for the servicing and repair of unmanned platforms and satellites, for the assembly and check-out of large space structures, for space-based scientific research, and for the development of new technologies.[22]

Try though he did, Gil Rye could not get the Department of Defense to buy these arguments. In fact, his colleagues at the Pentagon were furious with him for helping NASA prepare its case. It was not an accident, said one high-ranking Air Force official, that Rye did not return to the Air Force after his tour of duty ended at the White House. His military career ended when he embraced the space station.

Rye was obliged, given his position on the staff of the National Security Council, to present the views of the Defense Department and those of the other agencies represented on the drafting team. Rye's report began with a history of the President's space policy, described the rise and decline of American spending for space over the past twenty years, analyzed the Soviet space program, and explained the efforts of Europe, Canada, and Japan to enter the space race. It plowed through a summary of the main issues in the John Hodge report.[23]

On those matters, Rye got the drafting team to agree. Halfway through the twenty-two-page single-spaced report, however, the representatives diverged. Following a short description of the three options and their possible future elements, the report devolved once again into a statement of positions. The "discussion" of each option, to the extent that it could be called that, serialized the arguments of NASA, the Defense Department, and the other participants for and against the various options.

Some officials in the Air Force were prepared simply to state their position and stand idly by, reasserting their inability to use NASA's space station but not trying to kill it outright.[24] Their position did not prevail. Defense Secretary Caspar Weinberger decided to oppose the space station, not only on the grounds that the military could not use it but also with the argument that the country could not afford it. On August 8,

1983, two days before the SIG (Space) meeting at which discussion of the space station had been scheduled, President Reagan heard Weinberger express his personal opposition to the proposal. The most powerful member of the President's cabinet, Weinberger had known Reagan since the Defense Secretary was a California state legislator and the President was his governor. As governor, Reagan had tapped Weinberger to be his Director of Finance. Weinberger could be acerbic, and Reagan trusted him.

Weinberger met with Reagan in the Situation Room in the basement of the West Wing of the White House. Weinberger called the meeting with Reagan ostensibly to present the President with an overall assessment of the relative strength of the Soviet Union and the United States in space. The directors of the Central Intelligence Agency and the Defense Intelligence Agency also were there, along with a representative from the Joint Chiefs of Staff and the most senior members of the President's personal staff. The subjects covered at the meeting, dealing primarily with the military uses of space and the prospects for the President's Strategic Defense Initiative, were highly classified.

Weinberger laid out his objections to the space station in his opening remarks. According to Hans Mark, who represented NASA at the meeting, Weinberger told Reagan that there was no good reason for going ahead with the space station NASA was proposing. Given the limits on the federal budget, Weinberger said, other projects had much higher priority. Weinberger conceded that a space station would be built some day, but argued that this was not the time to do it. There were just too many other things that were more important. "I took issue with some of the things that were said," Mark remembered, "but I was clearly outgunned."[25]

The members of the Senior Interagency Group for Space met two days later, on Wednesday, August 10. Jim Beggs represented NASA. Rye presented his report. The representatives from the Joint Chiefs of Staff, the Central Intelligence Agency, and the Department of State all took positions against the proposal. Weinberger's presence at the meeting made an unusually strong statement against the initiative inasmuch as he did not hold a seat on the Senior Interagency Group for Space. His deputy had been named by the President to represent the Office of the Secretary of Defense, but Weinberger attended anyway to re-emphasize his opposition. "I was for it and almost everybody else was against it," Beggs complained. The group could not agree on any recommendation to take to the President. "It was clear that the Department of Defense was going to oppose this, and it was a straightforward decision on their part."[26]

Beggs returned to NASA headquarters, where he and Hans Mark, using the connections Mark had made as Secretary of the Air Force under the previous administration, tried to break the Defense Department loose from its position. Mark met with the new Secretary of the Air Force, Verne Orr, and the Director of the Defense Advanced Research Projects Agency, Robert Cooper. Beggs lobbied Weinberger, Orr, Pete Aldridge (who was Undersecretary of the Air Force at that time), and several high-ranking Air Force officers. "We believe we will be serving the national interest in providing an important research and development resource for the DOD," Beggs wrote Weinberger. "Military-related research and development" was the one area where the space station offered "the most efficient use of DOD resources," the Air Force team had admitted. Beggs pointed this out to Weinberger. "The Space Station will provide the opportunity to do research and to explore the potential military uses of a permanently manned facility."[27]

"My reservations about your proposal," Weinberger later wrote in his official reply, "relate to cost and impact on the Space Transportation System." In Weinberger's view, NASA's estimate of the initial construction cost of the space station "represents only a fraction of the actual costs required to achieve the initial capabilities you desire. Modules to make it operationally useful, and an extensive complement of instruments to support scientific missions, would inevitably multiply the total costs several times."

"In today's constrained fiscal environment," Weinberger noted, "unprogrammed cost growths can only be funded at the expense of other programs. I have continuing concerns about the ability of the nation to support and sustain major commitments to defense programs, as well as proposals like the President's Strategic Defense. You would not wish to cancel any of your approved civil programs to meet increased funding requirements for space station any more than we in Defense would like to see our national security budget jeopardized."[28]

"We remain firmly committed to the Space Transportation System," Weinberger insisted. His Air Force study group had recommended enhancement of the shuttle—longer stays in space, sophisticated satellite-retrieval systems, and "spacelab" type facilities—as an alternative to NASA's space station.[29] "We have reconfigured all our payloads to be Shuttle compatible and have invested a considerable portion of our space related funding in Shuttle related projects," he reminded Beggs.

"I believe that a major new start of this magnitude," Weinberger continued, "would inevitably divert NASA managerial talent and resources from the priority task of making the Space Transportation System fully operational and cost effective. With *all* our national security space programs committed to the Shuttle and dependent on it for their sole access to space, I am sure that you can appreciate my concern in this area."[30]

"We are very sensitive to the DOD concerns about the Shuttle program," Beggs assured Weinberger. "Getting the Shuttle fully operational and cost-effective is our highest priority." The space station, Beggs emphasized, would not "achieve initial operational capability" until "the 10th anniversary of the first Shuttle flight."[31]

"NASA is ready to begin the development of a space station," agency officials argued. "The Shuttle, always envisioned as a precursor to a space station, is now available. The technology requirements are understood and research is under way." Approval of the space station, NASA argued, "is not premature, but will provide an important resource for the national security community." If they did not need it at present, Beggs told Defense Department executives, they would surely need it later. The space station "would provide a facility in being for future national security activities at such time that the national security community develops requirements for a manned presence with unlimited stay-time in orbit."[32]

"They were not moved by our arguments," Beggs admitted. Accordingly, Beggs turned to John Hodge and the other members of the Space Station Task Force and told them to put the other rabbit in the hat. "As a result of that," Beggs said, "I told the study group we had put together here in Washington to forget the politics, write it on up, and develop our arguments for the space station with no military justification whatsoever." NASA, he decided, would ask the President to approve "a completely civil station," a facility that would serve only civilian purposes with no military missions.[33]

Weinberger was unimpressed. "I regret not being able to endorse the modified thrust of the proposed space station," he wrote Beggs, "but the national security implications are too extensive and are not mitigated by calling it a civil program."[34]

"Keep in the idea of international participation," Beggs told the members of the Task Force. Without the military on board, international cooperation would be much easier to achieve.[35]

18

The Number

As NASA officials searched for a method to put the space station proposal before President Reagan, they came under increasing pressure to make a public commitment on a very sticky issue. They came under pressure to tell the parties to the decision precisely what a space station would cost.

To some people, calculating the cost of the space station seemed like a fairly straightforward proposition. Identify the various components of the space station and estimate the cost of building them. Add on the funds needed to get the space station launched and running and produce what government officials called the "total program cost" for the facility.

In reality, the cost issue was much more complicated. Before they could estimate the cost of the space station, NASA officials had to make fundamental decisions about station design and program management. Task Force officials, however, did not want to design the station until the program was approved, and people within NASA were deeply divided over how to manage the program. These were tough decisions, and it was difficult to resolve them during the fragile approval process.

Up to that point in time, NASA officials had managed to avoid a specific commitment on the question of cost. A specific commitment at the start of a new technology program, they insisted, would be premature. Such a commitment might lock NASA onto a target it could not meet, an experience the agency had painfully endured when it promised to build the space shuttle for considerably less than the program actually cost. A specific number might also give opponents of the space station a bone on which to gnaw, a means by which to force negotiations on the design and scale of the project before it was even approved.

The space station, James Beggs said on more than one occasion, was something the President could purchase "by the yard." The initial cost of the facility would vary depending upon the number of modules the government wanted to install on the initial facility. It would vary with the size of the crew. It would vary according to the number and type of experiments the government wanted to conduct and the electric power needed to support them. If the President was willing to encourage international participation, foreign partners might share in the cost of the space station, adding to its value and complexity. And the space station would grow. After building the first phase of the space station, NASA officials wanted to increase the size of the crew, add more modules, and expand the electric power generating plant. That would cost more money. Given all of these factors, a specific number could be quite misleading. As an evolving facility, the only point in time at which the space station would cost a specific amount would be the day before it cost $1 million more.[1]

Task Force leaders did not want to commit NASA to a specific figure until the agency had completed its "phase B" definition and preliminary design studies—some two years after the President approved the program. NASA engineers would use the phase B studies to design the space station and define exactly what sort of technology they would need to build it. Only then, the leaders of the Task Force argued, could NASA come up with a precise cost estimate. If the cost estimate from the phase B studies came out too high, NASA could reduce the size of the space station "by the yard." In the meantime, Task Force leaders talked only about "cost bands." Cost bands were ranges of money within which the Task Force thought the price of the space station would probably fall.

This attitude fueled the arguments of the opposition. It was absurd, they told NASA officials, to expect the President to review an undertaking of this magnitude without being told what the whole program would cost. Only by revealing the total program cost, the opposition argued, could NASA hope to get the program approved. Only by getting the program approved, the Task Force leaders replied, could NASA start the process needed to determine the cost.

Given NASA's unwillingness to stand by a specific number, Weinberger and Stockman argued that the price of the fully developed space station must be astronomical. NASA must be trying to hide the total program cost. They also suggested, given the reluctance of NASA officials to put their finger on a specific number, that NASA executives did not know exactly what the space station would cost.

To a certain extent, this was true. Until Luther Powell arrived in Washington in April 1983, the Space Station Task Force had not conducted any serious cost studies on the overall facility. Some people thought that a bare-bones space station could be built for less than two billion dollars. When the Task Force was formed, John Hodge had assumed that a workable space station could be built for a cost band of $3–$5 billion. Those numbers remained on the "top twenty" viewgraphs for about two months. "I cried on my chair," Powell said as he recalled his encounter with the lowest of the cost estimates. Well before Powell's Concept Development Group assembled in Washington, Hodge had moved the Task Force estimate into the $4- to $6-billion range. "I knew we couldn't do anything for that," Powell responded, having conducted a preliminary analysis. "That was just out of the question. We could put an unmanned satellite up there, but you couldn't put a fully manned space station."[2]

Powell took his initial estimates to Hodge, Culbertson, and Beggs, who made another step increase. Hodge changed the viewgraphs so that by May 1983 they showed a $7.5- to $9.0-billion cost band. Task Force leaders thought that they could fit a technically capable space station within a $7.5- to $9.0-billion cost band, although they knew it would be tight.[3]

With the cost estimates on an upward march, opponents of the space station grew even more suspicious. Powell ran an open-door operation, with "a lot of contractors walking in and out." Somehow word leaked out that the Concept Development Group was analyzing options that reached as high as $12 billion. "They saw these numbers going anywhere from $7 up to $12 billion," Powell observed. "Somehow they got to chide back to Beggs, 'You guys don't know what this space station costs, you don't know what's in it.'"[4]

As the space station initiative moved toward the President's desk, the desire of White House officials for a good look at a specific number began to outweigh the efforts of NASA officials to avoid one. In a series of meetings, NASA executives began to lean toward an $8-billion mark. "It just emerged gradually," said Philip Culbertson. For Hodge, $8 billion was the highest number for which he could ask within the current cost band without setting off a disabling reaction. He observed how people outside NASA reacted to the various cost bands he had presented over the past year. "I reached the scream level at about $9 billion," Hodge recalled.[5]

Beggs leaned toward the $8-billion figure because of the mathematics of budgetary politics. He believed that the President would allow the NASA budget to grow a "little bit": the current budget adjusted for inflation plus a "little bit of growth." An $8-billion space station would fit under that budget curve, given the ups and downs of other NASA programs. "If you picked a number too high," Beggs explained, "OMB would come back and argue that we were busting the budget in the out years. And if you picked too low a number, you knew you would be in the soup very quickly."[6]

From his perch in the White House, presidential counselor Edwin Meese concurred with NASA's space station cost estimate. "Let's get our foot in the water," he agreed, "so that we have a commitment and then we can worry about the long-range costs later." If NASA had come in with a $12- to $13-billion space station, Meese observed, it "probably would have scared off the OMB people and might have made it less feasible."[7]

Even as they leaned toward the $8-billion mark, NASA officials continued to explore other options. "There were arguments for bigger, smaller, everything else," Culbertson explained.[8] Although Task Force officials revealed the $8-billion estimate to White House officials and representatives from other agencies during the SIG (Space) study process in the summer of 1983, they continued through October of that year to display a "top twenty" chart that showed the space station in the $7.5-billion to $9-billion range.[9] Officially, NASA's leaders maintained that the cost of the space station was still being studied. "An approved configuration does not exist and some of these elements as well as estimated costs could change."[10]

Having seen the $8-billion mark, people began to question it. The figure seemed suspiciously low. Other estimates ranged up to $14 billion and eventually on to $30 billion.[11] How could NASA even come close to an $8-billion limit? The answer, not untypical for a large-scale technology program of this sort, had more to do with the approach the Task Force took to the program than with the group's skill at making technical cost estimates. The Task Force approach contained some rather controversial guidelines about how the program should be managed.

The basic estimate for the space station itself was not much in dispute. The main space station, as the Task Force defined it in the summer of 1983, would hold a crew of six to eight astronauts. It would contain a living quarters module for the crew, two laboratory modules, a utility module, and a docking hub. It would possess a modest capability for servicing and repairing satellites. In addition to the main space station, program funds would also pay for two unmanned platforms.[12] Luther Powell's Concept Development Group calculated the equipment necessary to operate a space station of this scale—the electric power, computer power, thermal control, attitude control, and the various modules and platforms. The members of the group estimated the cost of building this equipment and added on the costs that the contractors would charge for engineering, design, testing, and overhead. Six billion dollars, they estimated, would cover the cost of developing the hardware. The exact amount might grow or shrink a bit, depending upon the sophistication of the station, but it was an estimate whose accuracy was not seriously challenged in the debates that followed.[13]

Estimating hardware costs is a technically challenging task. It was not, however, the crux of the cost problem. To deliver an $8-billion space station whose design and construction costs already exceeded the $6-billion mark, Task Force leaders had to make some very strict assumptions about the rest of the program. Some of those assumptions turned out to be highly controversial.

NASA officials typically included a cost "reserve" in their estimates to cover the changes that inevitably showed up in a project of this kind. Some people, especially those at the NASA field centers that would manage the program, wanted a big cost reserve. The Task Force leaders directing the fight for the space station, however, argued for a small reserve. A big cost reserve, three times larger than the small cost reserve, would add nearly $3 billion to the cost of the space station.[14] During the effort to get the space station approved, proponents of the small cost reserve prevailed. "The more you put in the budget the more they will spend for the same capability," John Hodge explained, talking about the people who would actually build the station. Philip Culbertson agreed. "It's just human nature in this game that if you have a 30 percent reserve you are going to spend the 30 percent reserve," he said. "I was willing to go into it with a tight reserve, because I felt that was the best chance to keep costs down."[15]

Within their $6-billion estimate of the cost of developing space station hardware, Task Force leaders included money that would be spent on the ground. Equipment would be needed to test components of the space station, train the astronauts, and monitor the facility once it was in orbit. Hodge wanted to keep these costs at a minimum, as much to change NASA's approach to human space flight as to save money. People at the NASA field centers, on the other hand, wanted to develop a "more comprehensive" system of "ground-based support infrastructure." They also wanted to use the space station, NASA's guiding program for the next decade, to augment their "basic engineering and technical support activities." The latter proved especially controversial. To make up for shortfalls in their budgets for administration, NASA field officers pulled money out of budgets for programs to finance their technical capabilities on the ground.[16]

Hodge did not want to divert a great deal of money into the construction of elaborate ground-support facilities for use once the space station was in orbit. He believed that the manned space flight program spent too much money on items other than hardware, and he wanted to change that practice. He did not want to pay what he characterized as the "institutional tariff," the funds taken from program budgets to pay for general technical capabilities. He was particularly adamant on that issue: "I kept pushing to say we are not going to put that tax, which is about 30 percent, in the cost of the program."[17]

The difference was significant. A more elaborate ground support system plus the "institutional tariff" could add nearly $4 billion to the $8-billion estimate. Hodge admitted that the more elaborate approach was "deeply rooted in the NASA culture." As long as he led the Space Station Task Force, the stricter philosophy seemed to prevail.[18]

Beyond the cost of constructing the space station, someone had to pay the cost of operating it once it was built. Someone had to pay the cost of the shuttle flights that would transport the components of the space station to orbit. Someone had to pay the cost of the science and technology experiments that would be conducted on the space station. Someone might have to pay the cost of a "lifeboat," a special landing vehicle that could return the crew to earth if an emergency occurred.

To someone unfamiliar with the NASA budget, "total" program costs meant exactly that—all of the costs required to build and operate the space station. Such a definition, however, could add $11 billion to the cost of the overall program.[19] Because of the way the NASA budget worked, NASA officials did not have to include these costs in the $8-billion estimate. Money spent on behalf of the space station by other program offices was carried on the accounts of those offices.

Launch costs would be financed through the shuttle program office. So could the free-flying robot. Instruments and experiments would be funded through NASA's offices for space science and advanced technology. Task Force officials decided to budget station operations separately from station development and to defer funds for a "lifeboat" until work on the space station design revealed whether it might be better to provide astronauts with a "safe haven" inside the station. Task Force officials repeatedly announced that these costs were not included within their $8-billion estimate.[20]

As the fall budget season approached, NASA officials grappled with the pressure to stand behind a single number. The space station concept was still very fluid. NASA officials were looking at different technology options, different configurations, different ways of managing the program. Many people within NASA wanted to issue an estimate above $8 billion. Even Hodge wanted a higher mark, something in the $9-billion range. Although the $8-billion estimate had been discussed with White House officials, Hodge observed, many people continued to treat it as "just a line in the sand," a point of discussion rather than a firm commitment.[21]

White House officials had seen the $8-billion figure, however. The number had been included in the options prepared for the meeting of the Senior Interagency Group for Space on August 10. If NASA executives came back with a higher number, just a month or so after discussing the $8-billion figure, it would suggest that the cost of the program was out of control prior to its approval. Any change at this point would send the wrong signal to the White House. "Eight billion stuck," said Hodge, even though NASA had not begun the detailed definition studies on which he wanted to base a precise estimate.[22]

On August 24, 1983, James Beggs sent a letter to White House Chief of Staff James Baker confirming NASA's intent to stand by the $8-billion estimate that had appeared in the SIG (Space) report. A more detailed explanation of program costs appeared two weeks later, in response to a request by OMB. Margaret Finarelli, NASA's representative on the SIG (Space) drafting team, sent a letter to the President's budget analysts, explaining what the $8 billion would buy and NASA's year-by-year schedule for spending it. Eight billion dollars would buy the living-quarters module, the two laboratory modules, the utility module, the docking hub, make a start on a modest satellite-servicing capability, and pay for two unmanned platforms. "The space station could be operational as soon as 1991," the letter read. "Because of our extensive planning efforts to date, we believe this program is as well estimated as any similar program at this stage of development."[23]

Commitment to the $8-billion mark caused considerable consternation within NASA. "That was as bad at that point as the four to six was when I first came," Luther Powell complained.[24] Not only did the $8-billion estimate sit at the lower end of the Task Force cost band, but it precipitated another disagreement over what the estimate included. All along, John Hodge and his colleagues had insisted that NASA spend three years and as much as 10 percent of the cost of studying the uses and requirements of the space station before developing it. Hodge wanted to add this 10 percent "definition" money to what he saw as $8 billion for "development." Such a move, however, would create the impression of cost growth, an impression that people in the NASA Administrator's office were determined to avoid. They decided to include definition costs in the $8-billion estimate, a commitment that was reaffirmed in the letter Margaret Finarelli sent to OMB. When NASA officials assigned $630 million for definition, they thereby left themselves less than $7.4 billion for development of the space station itself. A $7.4-billion expenditure for space station development actually fell below the lowest point on the current Task Force cost band. John Hodge and Philip Culbertson were quite upset with the way this decision emerged.[25]

"From that point on, we really had to do some scrubbing," Powell explained. With the number officially on the table, White House budget analysts began to scrutinize the estimate. "You couldn't come up there and just wave your arms or take a slice off of this and a slice off of that with no justification, so it really made us sharpen our pencils and sharpen our wits." Task Force officials worked to meet their estimate by eliminating technology options, squeezing cost reserves, restraining the expenses of ground support, and resisting growth. "I can't afford to buy one. That was my theme to everything anybody brought me," Powell said. "If there is something here that you will replace with what you want, we will do that, but when it got to that point where we were struggling with the cost of it, I just couldn't let anybody bring anything in new."[26]

In fact, on paper the space station continued to grow. NASA added two logistics modules and upgraded the power requirements from 60 to 75 kilowatts. One of the two laboratory modules mentioned in the Finarelli letter was taken off, but two new laboratory modules contributed by the Europeans and Japanese were added on. In spite of their assurances that the space station could be adjusted "by the yard," Task Force officials made no significant reductions in the major components of the space station in order to meet the $8-billion estimate.[27]

With the $8-billion cost estimate in place, NASA officials were ready for the final phase of the White House review process. They were ready to put the proposal on the President's desk, if only they could find an appropriate way.

19
Reagan

The President's budget director, David Stockman, opposed NASA's proposal for a permanently occupied space station. The President's science adviser, George Keyworth, opposed it. Ronald Reagan's long-time associate and Secretary of Defense, Caspar Weinberger, opposed it and kept insisting that NASA cite no military justification for the project. NASA executives had tried to move the initiative through the budget review process in 1982 and failed. By the end of the summer of 1983, it had become apparent that the Senior Interagency Group for Space would not send a favorable recommendation to the President, despite the fact that Gil Rye, the Air Force colonel who served as secretary to the group, was busily promoting it.

At the same time, President Reagan had repeatedly stressed his desire for a stronger space program. He seemed willing to make a decision on the next step for the American space program, even though the members of his administration disagreed on what that step should be.

As governor of California from 1967 to 1975, Reagan had watched helplessly as the U.S. Congress reduced NASA's budget from $5 billion to $3 billion. "The only time you can get a quorum in the Senate these days," he complained in remarks prepared for delivery at a Republican Party dinner in San Diego in 1971, "is when the Democrats fly back to Washington to vote against an aerospace appropriation."[1]

"The United States," he told the members of the California State American Legion at its 1971 convention in Los Angeles, "[must] maintain the technical capacity that gave us the world's highest standard of living and which may one day be called upon to produce the productive miracles that could assure national survival" (a not too oblique reference to the use of aerospace technology for national defense). "Many of our own planners," he added, "believe that those who control space may hold an unbeatable military advantage."

"Two weeks ago," he noted in the 1971 speech, "the Soviets put into orbit what is purported to be the world's first manned space laboratory. They have two rockets racing our own Mariner to Mars and there is speculation that theirs may attempt a landing on that distant planet."

"Today we hear a chorus counseling retreat—turning away from the next great frontier," Reagan warned. "They are not willing to make the investment necessary to keep America moving forward, to keep the greatest scientific teams ever assembled working on productive programs that will provide lasting benefits for mankind." A healthy aerospace industry "has kept America ahead—in first place—in the technology a modern nation must have to compete and even to survive."[2] Some day, Reagan predicted in a 1970 speech, "the many uses of space technology will make our investment in space as big a bargain as that voyage of Columbus which cost $7,000—and which was denounced as a foolish extravagance."[3]

Although Reagan believed that the U.S. lead in space had slipped badly, he did not make the space race an issue in the 1980 presidential campaign. He did not mention a space station in any of his campaign speeches.[4]

Reagan took office with a general desire for a more aggressive space program. Beggs came in with a specific commitment to a space station. To translate the President's general desire into something specific, Beggs needed to demonstrate that people outside of NASA supported the space station. Unwilling to bargain away the goal of a permanently occupied station as a means of winning the support of skeptics, Beggs tried a different tactic. He and his allies sought to maneuver the issue into a different forum within the White House, where supporters were more numerous.

As soon as he became NASA Administrator, Beggs approached presidential assistant Edwin Meese for help. More than any other adviser in the White House, Meese understood how the President would react to specific pleas. "The President has always had a very good historical perspective on things," said Meese, who had served as Reagan's Chief of Staff during the gubernatorial years and stayed on as his chief policy adviser through the national campaigns and into the presidency. "There is a little anecdote," Meese remembered. "When he [Reagan] was governor he had a meeting—this was during the campus unrest—and he had a meeting with all the student body presidents or other representatives of various California campuses. And one of the kids, I think it was a girl, challenged him."

"You don't know what its like to live in a generation that has television, instant communication, and transcontinental flights in less than six hours and space travel," she said, challenging Reagan's ability to understand the modern generation.

"And the governor said," Meese recalled, "No, our generation didn't have these things when we were your age. We invented them."

"That," Meese said, remembering the story, "epitomizes his historical view and appreciation of what had happened in his lifetime in terms of technological change and his commitment then as a governor and now as president to maintain that edge as far as the United States is concerned and that really is at the origin of his interest in space." Ronald Reagan has "always had an interest in space," Meese observed. "That interest was constant throughout the presidency and from even before."[5]

Meese became an early supporter of the space station idea. "It came primarily out of the policy work that I did during the 1980 campaign," he remembered. Meese had served as the campaign Chief of Staff, Reagan's senior issues adviser, and, after the election, as head of the transition team. Although he had worked briefly in the California aerospace industry during the 1970s, his first full exposure to the space station idea did not occur until the 1980 election. "We had a variety of advisory groups and talked with a lot of people during that time as to what the President's overall set of objectives ought to be for the future." Meese gave Beggs "the primary credit" for focusing attention on the space station within the administration. "The focusing on the space station as the next major project was an evolving thing," Meese remarked. "The President was certainly interested in the idea, but I mean he did not wake up one day and say, 'Let's have a space station.' "

As for Meese's statement on national television in 1982 that the administration was supportive of the space station program, it "was more my own speculation of what I felt," he recalled. The President, Meese suspected, was in no position to consider a space station that year, because of the recession and the size of the deficit. "It may well have been a budgetary caution from the OMB people that asked the President not to get too far committed," Meese reasoned, remembering why Reagan did not mention a space station during his 1982 Independence Day speech at Edwards Air Force Base.[6]

Meese and his assistant Craig Fuller worked to move the space station into a more favorable forum than the Senior Interagency Group for Space, where opposition from the defense community remained strong. Under the SIG (Space) umbrella, John Hodge's working group had bogged down in bureaucratic politics, while Gil Rye's drafting team had failed to reach any consensus. Interest in the commercial fallout from space programs gave Meese and Fuller the opportunity to maneuver the issue into a new forum where other voices could be heard. In the summer of 1983, Meese and Fuller arranged a meeting at which Reagan could discuss commercial opportunities in space with the heads of eleven U.S. corporations. Beggs worked on the invitation list. "He was very good in the briefings," Meese said of Beggs, "being in contact with the private sector people and knowing who to bring in." The executives met with Reagan for lunch on August 3, 1983.[7]

As a two-term governor of California, Reagan had gained plenty of firsthand experience with the commercial fallout from the space program. He had served as chief executive of a state that received four times as much money in NASA prime contracts as any other state in the union, including the prime contract for building the space shuttle. As governor, he had made speeches praising the economic and technological "spin-offs" from space exploration. The corporate executives, however, wanted to discuss more than spin-offs on the earth. They wanted the federal government to promote the commercial exploitation of space itself. Private industry was prepared to invest in space, in high-technology operations like zero-gravity manufacturing and privately owned launch services, if only the President would remove government policies inhibiting these developments. Reagan was clearly interested.

NASA had already done a great deal to stimulate interest in space commercialization, forming task forces, holding conferences, and sponsoring experiments on manufacturing and materials-processing in space. Proponents of space commercialization used arguments like those that had propelled European investors to put money into the American frontier. It was an old argument, holding out the promise of a new industrial revolution if only the government would establish policies encouraging private investment and subsidizing the effort in various ways.[8]

The business executives told Reagan that the federal government did not have a coherent policy for promoting space commercialization. They surprised him by criticizing his space shuttle policy, commenting on the way in which the government scheduled commercial payloads on shuttle flights and determined the price for launching them. They asked Reagan to establish a stable, consistent government policy promoting space commercialization which their companies could use as a basis for making their own long-range investment plans. One thing, more than anything else, they told the President, would give coherence to the government's efforts to promote the commercialization of space. What they wanted, they said, was a space station.[9]

Of the eleven executives sitting around the table, four headed companies that had helped write NASA's mission analysis studies and might expect to win contracts to build the space station if Reagan approved it. Most of the executives, however, held no direct interest in the space station. Some wanted to compete with NASA to build launch vehicles. Others wanted to build their own orbital platforms. They were prepared, under the right conditions, to invest in space.

Without signing any pledges and while qualifying his statement, according to one report, "by commenting on the realities of federal budgeting," President Reagan expressed his interest in a space station too.[10]

Gil Rye got the same reaction one week later when he briefed Reagan on the outcome of the August 10, 1983, meeting of the Senior Interagency Group for Space. Without making any verbal commitments, Reagan gave what Rye took to be a clear signal of interest in the space station initiative.[11]

Reagan's general expression of support did not deter the opposition. Keyworth, Stockman, and Weinberger continued to lobby against the space station. Stockman and Weinberger warned about the cost of the space station and the effect it would have on other programs—especially the space shuttle. Keyworth, in an effort to excite NASA's ultimate vision of space exploration, kept urging NASA officials to ask for something more. Something more adventurous, such as a base on the moon, would help fulfill NASA's desire for a long-range goal, but it would also be more expensive. NASA Administrator Jim Beggs believed that he had read the President's sentiments correctly. Beggs believed that President Reagan, though a devotee of space exploration, would not approve much more than the $8-billion space station that NASA had proposed. To the consternation of NASA officials promoting the space station, however, President Reagan proceeded to deliver a speech that seemed to embrace Keyworth's position. On October 19, 1983, NASA celebrated its twenty-fifth anniversary as a government agency by holding a party at the National Air and Space Museum, with Ronald Reagan as the featured speaker. "NASA's greatest gifts have been the moments of greatness that you've allowed all of us to share," Reagan told Jim Beggs, shuttle astronauts, NASA employees, and other guests. "NASA's done so much to galvanize our spirit as a people, to reassure us of our greatness and of our potential."

"Right now we're putting together a national space strategy that will establish our priorities, guide and inspire our efforts in space for the next 25 years and beyond," the President reported. "We're not just concerned about the next logical step," he said, making a not-too-subtle reference to the theme Beggs had selected to promote the space station initiative. "On this 25th anniversary, I would challenge you at NASA and the rest of America's space community: Let us aim for goals that will carry us well into the next century."[12]

Reagan seemed to be stating a preference for a non-incremental goal. This placed NASA executives in the curious position of having to assume that Reagan was not speaking for himself. Hans Mark, Deputy Administrator of NASA, learned that George Keyworth's office had drafted the speech. Beggs and Hodge viewed Keyworth's challenge as a red herring, a tactic to divert NASA from its space station goal. Something more adventurous, Hodge believed, like a lunar base, would cost at least $45 billion and take twenty years to build. It would cost so much and be so far off that the President would not approve it.[13]

NASA officials stuck to the $8-billion space station as their most expensive new initiative. Gil Rye's drafting group had rewritten the original options so as to turn the issue into an up-or-down decision on that initiative. The alternatives to the $8-billion space station, Rye's group wrote, were deferral of the issue or the evolutionary development of the Space Transportation System, which was the same thing as deferral. Although Keyworth continued to argue for a more adventuresome initiative, no such option was officially put before the President.

Rye, Beggs, Meese, and Fuller searched for the best means of putting the space station proposal before the President for a decision. Defense Department opposition ruled out SIG (Space) or the National Security Council. Proponents needed a broader forum, one in which issues that went beyond the national security implications of the space station could be heard. They found such a body in the Cabinet Council on Commerce and Trade. Like the National Security Council, this was one of the minicabinets through which presidential business flowed. Its membership, however, was much broader, including representatives from the domestic departments as well as the national security community. Malcolm Baldrige, the Secretary of Commerce, chaired the council. The name of the council emphasized the commercial potential of space to which Reagan had been exposed.[14]

Meese and Fuller scheduled a meeting of the Cabinet Council for December 1, 1983. Beggs and Mark prepared for the meeting with the help of the Task Force leaders, with Peggy Finarelli, and with Norman Terrell, who as NASA's Associate Administrator for Policy had advised John Hodge on how to run the SIG (Space) working group. The Task Force got the model-builders at NASA's Langley Research Center to build an impressive 5-foot-wide mock-up of the space station. Terence Finn prepared eleven viewgraphs, all new, for Beggs to use. Beggs lobbied the meeting's attendees.

Finn, Hans Mark, and John Hodge piled into a NASA station wagon to take the model to the White House. Once inside the Cabinet Room, they practiced with the lights to make sure the viewgraphs could be seen. "We put the model right by the door," Finn said, where the President would come in. As the cabinet members arrived, "Mark would snag people and show them this and it had a little astronaut on the shuttle and it was really beautiful."

"And when the President walked in, Hans Mark, fearless man that he is, got up and explained the model to the President," Finn explained. "Whatever this model cost, it was worth it."[15]

Baldrige opened the meeting and Gil Rye called everyone's attention to the twenty-two-page SIG (Space) interagency review paper, which had been distributed in advance but probably not widely read. Baldrige then turned to Beggs, who began his presentation with quotations from Presidents Kennedy and Nixon committing the United States to lead the voyage into space.[16] Over a picture of the Columbia landing, words announced that the space shuttle had provided the United States with "leadership in space for the 1980's." The space station would provide "a highly visible symbol of U.S. strength." It would stimulate advanced technologies, scientific and technical education, and commercial endeavors in space, the fifth and sixth viewgraphs read. The United States needed men in space, Beggs continued, not only to work with instruments and maintain and repair satellites, but also to react to the unexpected and capture the imagination of the public. Pictures of the space shuttle and space station accompanied these phrases.

The next viewgraph showed a picture of the Soviet Salyut station in space, followed by a schematic diagram with the details of the Soviet station described in Russian. Beggs then reviewed the options before the President, essentially in the same words as those contained in the Rye report. The President could commit to a space station now, he could expand the capabilities of the space shuttle and build unmanned platforms, or he could defer the decision.

Beggs ended with a single viewgraph that showed an American on the moon, the U.S. flag clearly visible on the astronaut's left shoulder, the space shuttle lifting off from the Kennedy Space Center, the rings of Saturn as seen from the Voyager flyby, and a photograph of the surface of Mars taken by the Viking lander. In the upper right-hand corner, above the words "Leadership in Space: the Next 25 Years," was a copy of a painting of the space shuttle docked to a small space station with the earth floating by.

Beggs turned to the President. The presentation, which took about 20 minutes, had been carefully designed to emphasize the space station as a means of securing America's leadership in space. As the most ardent supporter of space exploration since President Kennedy, Reagan was clearly committed to that objective. Beggs had not, however, convinced everyone in the room that the space station was the best means by which to achieve it.

Reagan asked for comments. Keyworth argued that the space station was an old idea and that an alternative like going back to the moon would be a better choice. Reagan asked Beggs what he thought of that. Beggs replied that a space station was a necessary step toward anything bigger.[17]

Paul Thayer, sitting in for Caspar Weinberger, stated the position of the Department of Defense, but hedged his opposition by saying that he personally liked the space station idea. David Stockman argued against the space station on fiscal grounds, as did the representative from the Treasury Department, R. Timothy McNamar.

"Somebody made a very bad statement," Beggs remembered, "and I started to respond. I looked at the President and he winked at me, so I bit my tongue and shut up."[18] Bill Brock, the special trade representative, gave the first really positive endorsement. Brock was for it, Beggs was for it, Baldrige was for it, Meese was for it, and Craig Fuller gave an impassioned plea on behalf of it. For the first time in the interagency review process, NASA executives were sitting in a room where the ayes at least equaled the nays. When David Stockman announced that the deficit would never go down if projects like the space station were approved, Attorney General William French Smith remarked that Queen Isabella must have heard the same story from her advisers. The President made a joke and everyone laughed.[19]

Reagan thanked the group and said that he would announce his decision soon. If he had been moved by NASA's plea, he did not show it to the crowd.[20]

The following day, Friday, December 2, 1983, Beggs met with David Stockman and the senior members of the White House staff to make the final decisions on NASA's budget proposal for fiscal year 1985. Beggs had originally asked for $225 million to begin work on the space station, an amount that had been whittled down to $150 million during the fall budget review. It was a tiny sum compared to the total cost of the facility. If Stockman agreed to the $150 million, he would initiate a space station program that would cost more than $8 billion down the road. To avoid that, Stockman and Khedouri had offered Beggs a compromise. They would give NASA enough money to start work on a $2-billion space platform that initially would be unmanned but could be expanded into a manned base later. If Beggs agreed, Stockman would support his request. If Beggs would compromise, they would take a single recommendation to the President. If Beggs refused, he would have to appeal the issue all the way to the President and take the chance that Reagan might turn the $8-billion space station down.

Beggs declined Stockman's offer. He would accept nothing less than an $8-billion space station with a $150-million appropriation for the next fiscal year. In that case, the budget board told Beggs, the issue would have to go to the President for a decision.[21]

In his presentation to the President at the Cabinet Council, Beggs and Finn had included a single viewgraph that showed NASA's budget from 1962 to 1983, plotted in constant dollars. It looked like a cross section of the Sierra Nevada viewed from the north: steeply rising from the plains, peaking off in 1965, then sliding down to a great central valley. It was meant to make the point that NASA had not received its fair share of federal tax revenues since the Apollo program days and that U.S. leadership in space had suffered as a result. To build an $8-billion space station,

Beggs needed a rising budget, which the chart projected cautiously.

David Stockman, guardian of the public purse, wanted to know how much of an increase NASA needed. If he could get away with it, Stockman wanted to hold NASA to no increase, forcing the agency to finance any space station by cutting back on other programs. Beggs felt that NASA could not build the space station without climbing out of the valley of spending into which it had fallen in previous years. It was for that reason that Stockman and Beggs squared off in front of the President in the Oval Office on Monday, December 5, 1983. Rye, Meese, and Craig Fuller sat in Beggs's corner, while Fred Khedouri backed up Stockman.[22]

Stockman asked Beggs if he was really committed to the numbers he had proposed. Beggs replied. "We had said, 'Eight billion dollars was the cost of space station,' but I went through my litany to the President, which I had done a number of times, that NASA had had a declining budget for all these years and if we were going to do this, we would really need a little bit of growth."

Stockman, Beggs remembered, offered to give NASA a level budget that would keep agency spending even with inflation. "And I kept saying, 'No, we need an appropriate increase.' Finally Stockman said, 'Well, we can concede one percent,' and the President said, 'Done!' " Both Stockman and Beggs apparently went away from the meeting thinking that they had won. Referring to the 1 percent real growth, Beggs heard Reagan say, "I can guarantee you that as long as I'm in town."[23]

Only a few people at NASA knew what Reagan was going to do. He had not made any public announcement and had told the people who did know to keep quiet. Beggs, who had learned of the President's decision the previous weekend between the two budget meetings, thought that the President would announce his decision with a few sentences in the budget message. Reagan, however, had a better idea. "It wasn't ours," Beggs said, referring to the President's plan for making his decision known. "I wish I'd been that smart, I would have recommended it."[24]

PART IV

20
Congress

"Mr. Speaker," Doorkeeper James T. Malloy cried on the twenty-fifth of January 1984, "the President of the United States." Ronald Reagan, the fortieth president under the nation's 197-year-old Constitution, worked his way toward the three-tiered podium of the House of Representatives where House Speaker Thomas P. O'Neill and Vice President George Bush waited. Walking down the center aisle, Reagan shook hands with congressional leaders and acknowledged the applause of the assembled ambassadors, justices of the Supreme Court, and members of his cabinet.

Outside the Capitol, police officers formed a loose line around the hill on which the legislative chambers stood. Other officers gathered in knots on the roads leading up to the marble halls. A police helicopter circled overhead, swinging its searchlight across the neighboring grounds. The State of the Union message was the one certain occasion on which nearly all of the officials in the line of succession gathered in a single place at one time. All attended save a single cabinet member, selected to stand by at a hidden location should a catastrophe befall the entire leadership of the nation.

The police had blocked off all of the major roads surrounding the Capitol building. Just a few hundred feet from the barricades, at a Holiday Inn across from the building housing NASA's top executives, members of the Space Station Task Force and many of the people with whom they had worked for twenty months gathered around television monitors placed in rooms named after the nineteenth-century explorers Meriwether Lewis and George Rogers Clark. The Task Force had grown to seventy-two participants, plus three dozen members of the Concept Development Group working full-time in Washington. Including the various working groups and steering committees, more than five hundred NASA employees had helped prepare the space station initiative, nearly five percent of the total number of engineers and scientists employed by the agency in Washington and the field. The Washington contingent of NASA employees was joined at the Holiday Inn by interested contractors, international contacts, congressional staffers, White House aides, and allies from other agencies with an interest in the initiative. There had been rumors that President Reagan might mention the space station that evening in his State of the Union address.[1]

"Members of the Congress," Speaker O'Neill announced, "I have the high privilege and distinct honor of presenting to you the President of the United States." Reagan stood on the second tier, behind the lectern normally occupied by the Chief Clerk of the House, acknowledging the applause. He planned to use the speech to set the theme that would guide him through his 1984 re-election campaign.[2]

"There is renewed energy and optimism throughout the land," Reagan read. "America is back, standing tall, looking to the eighties with courage, confidence, and hope."[3]

"It's time to move forward again," he said. "Let us unite tonight behind four great goals." The first goal, he explained, was economic growth. He spoke for ten minutes about the need for a balanced budget. At the Holiday Inn, a few blocks away, only a few of the NASA employees and their invitees knew what Reagan's "second great goal" would be.

"A sparkling economy," Reagan continued, "spurs initiatives, sunrise industries, and makes older ones more competitive. Nowhere is this more important than our next frontier: space. Nowhere do we so effectively demonstrate our technological leadership and ability to make life better on Earth. The Space Age is barely a quarter of a century old. But already we've pushed civilization forward with our advances in science and technology. Opportunities and jobs will multiply as we cross new thresholds of knowledge and reach deeper into the unknown."

"America has always been greatest when we dared to be great," Reagan said. "We can reach for greatness again. We can follow our dreams to distant stars, living and working in space for peaceful, economic, and scientific gain. Tonight I am directing NASA to develop a permanently manned space station and to do it within a decade."[4]

In the hotel room four blocks away, Reagan's next words could not be heard over the cheering of the crowd. Some of the people in the room had been waiting for two decades to hear a president speak that line. "The euphoria was as high as I could recall for a long time," said Bob Freitag, Deputy Director of the Space Station Task Force, "comparable to that surrounding a successful key launch or recovery."[5]

Those who listened closely could hear Reagan praise the potential for "research in science, communications, in metals, and in lifesaving medicines which could be manufactured only in space." They heard him invite America's allies to join in the endeavor. "We want our friends to help us meet these challenges and share in their benefits. NASA will invite other countries to participate so we can strengthen peace, build prosperity, and expand freedom for all who share our goals." The statement on international participation surprised Robert Freitag, who had reviewed milder drafts of the speech as they worked their way up to the White House.[6]

When President Kennedy stood at the same lectern twenty-three years earlier and proposed that Americans race the Soviets to the moon, he made it clear that "this is a judgement which the members of the Congress must finally make." Worried that Congress might "reduce our sights in the face of difficulties" or choose "to go only halfway," Kennedy deviated extensively from his prepared text to rally the members of Congress behind his proposed initiative. He spoke directly to "the leadership of the space committees of the Congress and the Appropriations Committees." He spoke in great detail about the funds Congress would have to approve if the nation were to meet the new space goals. "I am confident," Kennedy told them, "you will consider the matter carefully. It is a most important decision that we make."[7]

Reagan's speech was much less condescending. "*I am directing* NASA to develop a permanently manned space station," he announced, "and to do it within a decade."[8] To head off any potential dissent on Capitol Hill, where the opposition party controlled the House and a continuing lack of consensus about space policy prevailed, NASA officials had made a number of moves. They had cultivated the members of four congressional committees that would have to authorize construction of the space station and appropriate the funds to finance it, briefing members and staff as the decision wound its way through the White House. NASA officials had promised to spend at least two more years studying the space station before asking for the first one billion dollars to build it. And they had asked for a very small first-year appropriation.

Ever since they had set the space program on an incremental path, NASA officials had been willing to accept smaller and smaller first-year appropriation requests as a device for getting big programs started. In approving the mission to the moon, President Kennedy asked Congress to appropriate $531 million in start-up costs—a huge sum by 1961 standards and a 50 percent increase over the $1.1-billion budget outgoing President Dwight Eisenhower had recommended for NASA. In 1972, President Nixon launched the Space Transportation System with a first-year appropriation request of $200 million, just 6 percent of the total NASA budget request that year. President Reagan initiated the space station program with a $150-million request—barely 2 percent of his overall recommendation for NASA—and a pledge to ask for no more than $250 million in the following year.[9]

The efforts of NASA officials to downplay the overall cost of the space station did not fool all of the lawmakers. "I will bet you a hat," Edward Boland said after inspecting NASA's overall cost estimates for the facility, "that it is going to cost a lot more than $8 billion."[10] A quiet, seventy-two-year-old New Englander, Boland had spent thirty-one consecutive years representing central Massachusetts in the U.S. House of Representatives. Elected to the Massachusetts state legislature at the age of twenty-four, Boland had spent a lifetime listening to administrative officials trying to justify their programs. He had even been an administrative official himself, serving for eleven years as the Register of Deeds in Hampden County, Massachusetts. With a sharp tongue and a quick mind, he was a very hard man to fool. He was also a Democrat, not a member of President Reagan's Republican Party. "I don't want to see us put up a space station," Boland said, "and spend $8 or $9 billion and then realize that we are two-thirds of the way down that track and don't have any money to put experiments on that station—or we don't have enough money to put up our own materials processing laboratory—or properly equip that laboratory." Boland worried that NASA had proposed a space station "that we can neither afford to operate or effectively use."[11]

Boland chaired the House Appropriations Subcommittee on HUD-Independent Agencies. His subcommittee reviewed spending requests for housing, urban development, environmental protection, veteran's affairs, disaster relief, and space. As one of the most senior members of the House, Boland involved himself in issues well beyond the jurisdiction of this broadly based committee. In 1982, he authored the famous Boland amendment that prohibited President Reagan from using government funds to support military forces attempting to overthrow the government of Nicaragua, an amendment whose inobservance nearly crushed the Reagan presidency. The space station initiative was only one item on Edward Boland's legislative agenda and neither the largest nor the most important to come before him.

"We don't have enough money to do all the things we would like to do," Boland observed. "It is just that simple." Roughly three-fourths of the federal budget consisted of uncontrollable expenditures, money that by law had to be spent on programs such as social security, food stamps, and unemployment insurance. "After all the uncontrollables are funded," Boland explained, "the remaining 25 percent comes under increasing pressure to make cuts. NASA is one of those agencies in that 25 percent." In order to make those cuts, Boland said, his committee could force all the affected agencies, including NASA, to take across-the-board reductions or it could set some "trade off in priorities between revenue sharing, toxic waste cleanup, NASA, and so on."

"In no program," he emphasized, "is that message more important than in the space station." The space and scientific community, he said, had "to come to grips with priorities."[12]

"How much of all this," Boland asked, referring to the missions the

Space Station Task Force had scheduled for the station, "could be accomplished by beginning first with the unmanned platform?" Boland argued that an unmanned platform "could include a number of free-flyer satellites for science missions and other automated functions, and then service the missions with an extended duration orbiter capability or man-tended approach."

"When we ask whether that approach wouldn't be cheaper—wouldn't it accomplish a significant portion of the potential NASA mission over the next 15 or 20 years—we are told you can't ask that question. That is not a question NASA wants to address."[13]

In the Senate, even though the Republicans controlled the chamber, the space station directive prompted similar inquiries. Senator Jake Garn, a strong supporter of the space program and chairman of the appropriations subcommittee reviewing the space station, wanted to know whether NASA could develop an automated facility, one that depended more on computers and robots than on people. An automated station, its advocates argued, could push America's genius for high technology toward new discoveries. In a letter to Craig Fuller, NASA Administrator James Beggs warned members of the White House staff that Garn, a member of the President's own party, "is advocating an increasing level of automation in the Space Station to the point of potentially delaying the introduction of the manned elements. The Space Station will, of course, be a highly automated system, and it will require many advances in automation techniques and robotics. It is, however, the presence of man which makes it a unique national resource." Beggs urged "that the President emphasize to the Senator the importance of putting momentum behind the program as a permanently manned facility."[14]

"Let's not put up a space station just so we can say we have one up there," Boland replied. "What we don't want to see happen is to put up a space station and have six people sitting up there in a habitat waiting for our materials processing laboratory—or a life sciences laboratory—that may be two years behind schedule because of funding problems." Boland went on to warn that "we are going to see some very tight funding for NASA and for every other agency in the HUD appropriations bill. So let's be sure that together we build something that we know is going to be of use to this country on the first day it becomes operational."[15]

Under normal circumstances, the President could turn to the ranking Republican on Boland's subcommittee and ask him to present the White House point of view. In this instance, however, that was totally out of the question. Bill Green, the fifty-four-year-old representative from the high-rent congressional district in Manhattan, New York, was the subcommittee's strongest opponent of a permanently occupied space station.

"Our space program has matured to the point where it no longer needs to depend exclusively on human daring," Green wrote. "The President's own science advisers are acknowledging that manned flight costs about five times as much as unmanned, and other scientists are suggesting that for many experiments man is either of no value or potentially disruptive." Green admitted that "our imaginations are not lifted by the vision of a robot reaching the stars," but he argued that if an automated space station visited only occasionally by human beings could "do the job as well and less expensively, we should take that path."[16]

Congressman Green represented a position held by large segments of the science community. Such concerns echoed through the national and specialized press. Tina Rosenberg, writing for *The New Republic,* called Reagan's decision a "mission out of control." Editorial writers for *Nature,* one of the oldest and most prestigious scientific journals in the world, called it "a tragedy." They bitterly denounced the project as "another two decades of original research on why astronauts vomit." *Newsweek* quoted planetary scientist Thomas Donahue, Chairman of the Space Science Board of the National Academy of Sciences: "If the decision to build a space station is political and social, we have no problem with that. But don't call it a scientific program."[17]

Editorial writers at the *New York Times* summed up the opposition in an editorial published four days after President Reagan's State of the Union address: "Far from being among the 'giant steps for mankind,' as Mr. Reagan suggested, it would be what his own science adviser called the proposal a year ago: a 'most unfortunate step backward.'" Almost every use of the space station, the *Times* writers asserted, "could be better accomplished without man. Telescopes can be pointed more accurately without humans lumbering around. Manufacturing in space hardly requires human presence." As for the argument "that humans must have a role in any space extravaganza if the public is to enjoy the show," the *Times* editorial staff argued that "an unmanned program offers far greater opportunities for stirring the public's imagination."

"Build vehicles that could roam over Mars—wondrously operated by a driver sitting on Earth," the editors continued. "Construct spacecraft from which earthbound viewers can feel themselves skimming the rings of Saturn, or sail over the clouds of Jupiter and watch its 16 moons rise and set. Wouldn't the public prefer that kind of spectacle to seeing another astronaut swing a golf club or cancel postage stamps on some distant piece of rock? Space is indeed a frontier. But there's a greater scientific payoff from putting human intelligence above human presence on space missions."[18]

Scientists were not only concerned that a multi-billion dollar space station would crowd out scores of unmanned projects, they also believed that the great discoveries waiting to be made in space would be made by robots. Robots could fly to Mars and back while the scientists were still alive; humans might not. Space telescopes could inspect distant galaxies and solar systems; human interstellar space travel seemed impossible. Most NASA engineers still held to the eighty-year-old vision of a permanently manned base on the edge of the space frontier; many scientists no longer shared that vision.

Even the members of the Senate Committee on Commerce, Science, and Transportation—who generally supported the space station they were being asked to authorize—called for a study of alternatives. In July 1982, seventeen months before Reagan issued his directive, committee members instructed their own Office of Technology Assessment "to assess the need for a permanent orbiting facility." Policy analysts in the Office of Technology Assessment (an arm of Congress) began to marshal arguments against the facility. As the lawmakers began to study President Reagan's directive, OTA analysts prepared their report.[19]

In the Republican-controlled Senate, the staff director for Jake Garn's appropriations subcommittee announced that he would convene a workshop to discuss an automated space station. With the assistance of officials in the Office of Technology Assessment, Wallace Berger wanted to examine "the antithesis of what NASA has charged for the space station and that is really asking what can be done without the permanent presence of man on orbit." He scheduled the workshop for March 1984, just as the various congressional committees would be finishing their hearings. Two dozen participants attended, mostly experts in the field of robotics. NASA's representatives (four in all) were outvoted. The day after the workshop, NASA executives learned that Berger intended to recommend that Garn's subcommittee reject the space station concept as NASA had proposed it in favor of an automated facility.[20]

Three times during the 1970s, Boland's subcommittee had deleted funds for major new starts on the grounds that NASA had not clearly established its budgetary priorities. In 1974 Boland and his colleagues voted to kill the initial development of the space telescope. In 1975 the subcommittee ordered NASA officials not to start work on the Pioneer probe to Venus. In 1977 the subcommittee blocked funds to start work on the Galileo probe to Jupiter. Each affected robotic programs, not the more glamorous manned initiatives. In each case the Senate appropriations subcommittee restored the funds to start the programs. These were signs, nonetheless, that the White House seal of approval on new NASA programs might be losing its value on Capitol Hill.[21]

Representative Boland went to the House floor in 1977 to argue for denial of funds for the Jupiter probe. "In the 20 years that this nation has been involved in a major space effort, the Congress has never, never made an attempt to deny funding for a major space planetary mission. I think it is time that we did." Boland urged the members of the House "to hold NASA's feet to the fire in determining what its priorities are." Boland accused NASA executives of avoiding choices by posing arguments about the next logical step. "Last year the 'next logical step' was the space telescope. This year it is the Jupiter Orbiter Probe," Boland had said in his unsuccessful attempt to get the legislators to stand firm against the Jupiter probe. "We are not stopping the Jupiter Orbiter Probe. We are not starting it. When we start these programs, we cannot stop them."

Boland had rescued the space shuttle when it ran into financial troubles in the late 1970s and, in his own words, "this subcommittee has dealt rather generously with NASA over the years." A permanently occupied space station, however, would cost an awful lot of money in the long run, and there was sentiment on Capitol Hill that Congress might want to put off any such commitment until the nation had more clearly defined its long-term goals for space exploration.[22]

The euphoria caused by the President's 1984 State of the Union address died down. Congress began to scrutinize the proposal. NASA's space station initiative, agency executives agreed, was in trouble again.

21
Momentum

Among the potential clientele for the U.S. space station, few were as potent as the international partners. The foreign allies rarely testified on Capitol Hill, and lawmakers only occasionally asked questions about them. Nonetheless, they were a very important factor in NASA's overall effort to win congressional consent for the President's decision.

Historically, Congress deferred to the President on matters affecting outer space. Space policy served as a symbol of U.S. leadership, a sign of technical and military prowess. Within the U.S. government, the responsibility for defining U.S. leadership among the nations of the world devolved upon the President. The responsibility for representing the United States in negotiations with foreign leaders also devolved upon the President. When a president initiated an agreement with the head of a foreign nation, U.S. lawmakers generally respected it. By involving U.S. allies so early in the development of the space station, NASA officials adroitly strengthened the President's prerogatives in space policy by adding the extra element of foreign affairs. To turn down the program, Congress not only had to say no to the President in his capacity as definer of space policy, but it also had to tell America's allies that the United States was pulling out of an international program just as the President in his capacity as chief diplomat was inviting them in.

To maintain international support, NASA officials had to convince their foreign counterparts that joining the crew of a U.S.-led space station would produce many benefits and few risks. This proved difficult. In Europe, Canada, and Japan, there was much debate over what sort of foreign cooperation might prove acceptable.

"As a follow on to his State of the Union Address," White House national security adviser Robert McFarlane wrote to NASA Administrator James Beggs, "the President would like for you to travel as soon as possible to appropriate foreign capitals as his personal emissary and meet with senior officials to discuss potential international cooperation on the U.S. Manned Space Station Program." Working with McFarlane, Beggs, and Ken Pedersen (the Director of NASA's International Affairs Division), Gil Rye had suggested that the White House use the space station issue to enliven the upcoming economic summit meeting with the leaders of Europe, Canada, and Japan. Rye briefed McFarlane and the two of them briefed Reagan and they all concurred. The summit was scheduled for June 1984, to be held in London just as the space station issue coincidentally came to a vote on Capitol Hill. McFarlane told Beggs in late February to develop "a framework for collaboration" with the allies, clear it with other U.S. agencies, and present it to President Reagan in time to be "formally proposed and agreed upon at the London Summit."[1]

Beggs left on his three-continent tour a few days later, spending much of March outside the United States. Most of the previous contacts between NASA and the foreign representatives had been at the working level, between engineers and scientists who shared a common commitment to space exploration. On his trip Beggs met with presidents, prime ministers, and members of their cabinets, discussions that took place at the highest political level.

Pravda editorials and news stories from TASS dogged Beggs as he traveled. Soviet leaders sought to link the space station program to President Reagan's "star wars" initiative, about which America's allies were quite nervous. "Washington demands from Tokyo large-scale participation in its plans of militarisation of outer space whose implementation Washington views as one of the main ways of achieving world supremacy," TASS commentator Aleksey Popov wrote as Beggs met with Japanese Prime Minister Yasuhiro Nakasone. "The Japanese press writes that the aim of his visit is to get Japan's consent to participate in the creation of a permanently manned orbital complex whose purpose is mainly to implement military tasks." Any such linkage would severely damage U.S. hopes for allied participation in the space station program. "The overwhelming majority of the Japanese and Europeans oppose the Reagan administration's plans to drag their countries into the arms race in space," Popov continued. "But the Japanese ruling circles disregard their own commitments to conduct space research in exclusively peaceful purposes and are apparently prepared to help Washington at its first bid in the implementation of militaristic plans in outer space."[2]

Prime Minister Nakasone reportedly told Beggs that Japan would participate in the space station program only if it was used for peaceful purposes. French President François Mitterrand told Beggs that there was no support in Europe for a military space station.[3] Had the U.S. Defense Department supported the space station, Beggs would have had a difficult time deflecting this concern. "During my trip," Beggs said, "I was also asked frequently about the extent of U.S. military involvement in the U.S. Space Station." He informed the allies that "while the Defense Department worked with NASA in the early planning for Space Station by reviewing their near- and long-term requirements for space, they concluded they had no requirements for a manned Space Station." The opposition of the Defense Department allowed Beggs to assure the allies that "the Space Station that the President directed NASA to build is a civil Space Station."

"Of course, like the Shuttle, the Space Station will be available for users," Beggs observed, qualifying his assurance in case the defense chiefs should change their minds. "If there are any national security users, like national and international users, they will be able to pay to use the facility." At the same time, Beggs assured the allies that "all activity on the Space Station will be limited to peaceful, non-aggressive functions."[4]

In the other nations, as in the United States, the officials proposing a major new human space flight initiative encountered disputes over space policy within governments having more projects than funds. Beggs asked the allies to contribute significant sums of money to construct and operate the station, money that otherwise might be spent on alternative space ventures. Lynette Wigbels of NASA's International Affairs Division, who was doing much of the staff work on foreign participation, estimated that the space station program might consume 25 percent of the space budgets of the European Space Agency (ESA), Canada, and Japan if they chose to come on board.[5]

In Canada, which had one of the smaller budgets, participation in the space station went up against unmanned priorities that in some ways provided a better return on the dollar than the more expensive human presence in space. The space station initiative in Canada competed with the proposal for a Canadian radar satellite—originally designed to provide all-weather monitoring of forest fires and sea lanes in the remote northern regions—and MSAT, a communications satellite system that would link mobile radio users in backwoods areas with the rest of Canada. These were very attractive projects and there was not enough money to fund them all.[6]

Like their American counterparts, Canadian civil servants promoting the space station had to build public and ministerial support for their own human presence in space. In some ways, this was harder in Canada than in the United States. Canada had no tradition of human space flight. So in 1982, shortly after being asked to participate in NASA's space station planning process, civil servants in the National Research Council of Canada decided to create one. They encouraged their government to organize a Canadian astronaut corps to fly on the American space shuttle.[7] Similar initiatives were already under way in Europe.

NASA lent support to the budding Canadian and European space flight programs in a special way. In 1983, NASA executives decided to fly the space shuttle Enterprise to Europe and Canada. It traveled abroad on the back of a specially adapted Boeing 747. Jim Beggs and Hans Mark led the tour, using the visit to promote international participation in the American space program. The reception in Europe, where governments were starting up their own astronaut corps, can only be compared to events during the early days of the U.S. space program. Mark, who was on board the 747, described the reception as the shuttle and its transport descended through the clouds above the British Isles:

"Since it looked like the weather was lifting, Joe Algranti (the pilot) called the Heathrow Tower and asked to fly down the Thames at 3,000 feet. Permission was granted and the next twenty minutes were absolutely fascinating. It was Sunday afternoon and many thousands of people were lining the river watching for the Enterprise to pass overhead. The crowds were enthusiastic—even at that altitude we could see them cheering and waving. Then, in the course of three minutes, we flew over the Parliament at Westminster, the Tower, and the famous observatory at Greenwich.

"If overflying London was an experience, then our landing at Stansted was completely overwhelming. Nowhere else in Europe did we get the kind of reception we experienced. There were crowds everywhere, but in this place the enthusiasm was hard to believe. The people cheered and held their hands up making the 'V' for victory sign and many had tears streaming down their faces. I cannot explain why this happened, I can only record it."[8]

The Enterprise was the main attraction at the Paris Air Show. On a return stop at the Canadian capital, a city with a population of 300,000, four hundred thousand people showed up to see it. At the welcoming ceremonies, the Canadian Minister of Science and Technology announced the creation of the Canadian astronaut program. More than 4,000 people applied. The National Research Council rushed to select six men and women so that one would be ready to fly on board the space shuttle Challenger by the fall of 1984.[9]

The first member of the European astronaut corps, German payload specialist Ulf Merbold, flew on board the space shuttle Columbia in late 1983. On December 5, 1983, NASA set up a conference call between U.S. President Ronald Reagan, German Chancellor Helmut Kohl, and the Columbia crew. On that same day Reagan met with Beggs and budget director David Stockman and officially approved the appropriation request for the space station. Enthusiasm for human space flight built steadily as NASA advanced its space station plans.

"The reaction so far to the President's call for international cooperation has been both strongly positive and openly appreciative," Beggs announced upon his return from Europe and Japan in 1984. "I heard nothing but praise for the President's foresight and leadership in making this decision." What he had not heard, however, was a willingness on the part of the allies to accept the President's invitation without first engaging in some really tough negotiations. Those negotiations would decide the terms and conditions of international participation, an area of persistent disagreement.[10]

"Each of our countries will consider carefully the generous and thoughtful invitation received from the President of the United States to other Summit countries to participate in the development of such a station by the United States," the foreign heads of state agreed. "Manned space stations are the kind of programme that provides a stimulus for technological development leading to strengthened economies and improved quality of life." It was not clear, however, whether other governments would launch their space flight programs "in the framework of national or international programmes."[11]

In other words, Europe and Japan could go it alone. The spacelab program had given the Europeans important knowledge about human space flight technology. Using that knowledge, the Europeans were preparing plans for the Columbus project, a free-flying space station module that could float independently around any low-earth-orbit American facility. The French were pushing ahead with plans for their own Hermes space shuttle. If ESA followed through on plans to upgrade its Ariane rocket into a heavy-lift launch vehicle, the Europeans could put Columbus into orbit from their own launch complex in French Guiana without American help. With the Hermes space shuttle, also to be launched by an Ariane V rocket, they could fly their own astronauts to and from their own low-earth-orbit facility.[12]

The Japanese had similar plans. They had designed a massive new spaceport on the Pacific island of Tanegashima from which they could launch a new H-2 heavy-lift vehicle. They were preparing to construct a free-flying space platform from which they could conduct ultra-precise space science experiments. Program planners also wanted to build a Japanese space shuttle that could be launched on top of the three-stage H-2 and carry Japanese astronauts into space.[13] In their space shuttle programs, the Japanese and Europeans were at least fifteen years behind the United States. In their technology for space platforms and modules they were fairly close, and in their planning for heavy-lift vehicles they were actually ahead of NASA, which had quit making large rockets like the Saturn V before the shuttle program began.[14]

The Europeans and Japanese had to weigh how much they would learn through cooperation with the United States on an international space station against going it alone. They had already endured one unpleasant experience over the technology transfer issue. Leaders of the Space Station Task Force had invited the international partners to attend meetings that NASA regularly held with its contractors to review space station missions and requirements. As work progressed, the Space Station Technology Steering Committee scheduled a conference to be held in Williamsburg, Virginia, in March 1983. The participants at this conference, both NASA employees and their industrial contractors, would select "a set of recommended advanced technologies which should be matured to support the initial space station." As usual, the leaders of the Task Force notified the foreign participants about the meeting. NASA higher-ups then withdrew the invitation, fearing that the presence of the international partners at a technology conference might provide ammunition to opponents of the program who worried that an international space station would lead to a massive hemorrhage of U.S. technology.[15]

NASA executives had to walk a very fine line in discussing the terms of international cooperation with the foreign governments. "Technology transfer has been an increasing concern on all our parts in the past few years," Beggs wrote to one of the international partners. They would receive "assured access" to the station, said NASA officials, but the program had to be "conducted to avoid the unwarranted transfer of technology among the partners."[16]

Running the space station so as to avoid technology transfer would not be an easy task. It would be easier if the United States allowed each nation to build its own part of the station and use it as each nation pleased, as the Europeans wanted to do with their detachable Columbus module. NASA officials, however, were strongly opposed to the notion that the Europeans might pick off their module and fly away. The space station would not be big enough nor would it have enough electrical power for every nation to hook up its own module and conduct its own experiments. Nations would have to share facilities. Because of strong congressional and presidential interest in the commercial potential of space, NASA would eventually insist that it be allowed to build the materials-processing lab. That would leave the Europeans with the less glamorous task of building the life sciences lab. To conduct materials-processing experiments, the Europeans would have to use a U.S. module. Furthermore, they could not just float in and use it. The experiments would have to be scheduled on the basis of international agreements acceptable to all of the partners and based on their relative contributions to the station.[17]

The Europeans continued to insist on a "genuine partnership," not one dominated by the United States. In one amusing episode, Beggs suggested that the space station be operated like a condominium, an apartment building in which all of the residents privately owned their own residential quarters but contributed jointly to the management of commonly held properties such as the swimming pool and tennis courts. Neither the Europeans nor the Japanese understood the term. The European Space Agency even went so far as to hire a Washington, D.C., legal firm to research condominium law.[18] NASA's hopes for international participation could fall apart on issues such as access and technology transfer. "The French will be tough bargainers," Beggs reported after his meetings in Paris in March 1984, "and obviously intend to pursue their own independent space programs." Other countries might be no less difficult.[19]

The leaders of the Task Force desperately wanted to keep the international component in the U.S. program. Their counterparts overseas were enthusiastic about it too. Rather than get hung up on the issues that divided them, NASA suggested that the U.S. and foreign participants approach the problem incrementally. Divisive issues, the Americans argued, could not be settled until each party knew exactly what it would contribute to the space station. By agreeing in general to cooperate, all parties could work together to develop a preliminary design for the space station and put off the tough, divisive issues until later.

The Europeans wanted to settle the tough issues before agreeing to join NASA in the "phase B" definition and preliminary design work. There was not time, NASA officials told them. The United States would start its phase B studies in June 1985, assuming that Congress approved the program. "The train is leaving the station," NASA told its partners. "We are going to do it anyway." If the international contingent did not get in on the phase B studies, the Americans would not be able to redesign the space station to put them on at a later date. The phase B studies would cost each partner a few million dollars, after which each could decide whether or not the terms of participation were acceptable.[20]

By putting off the really tough issues, allowing each party to retain its own view of the program, NASA officials could maintain the momentum that had already brought them so far.[21]

22
Management

In building support for the space station program, NASA executives had to maintain an additional, less glamorous type of momentum. They had to maintain enthusiasm for the project among their own employees. The program planners at NASA headquarters realized that they would be unable to do this unless they were able to tell the people at the field centers what their role in the new program would be.

NASA is an organization of field centers, fewer than a dozen campus-like installations where the work of the agency is done. Only 7 percent of NASA employees work at agency headquarters in Washington, D.C. (which the field employees prefer to view as the Washington center rather than the pinnacle of a hierarchy). The field centers have always been semi-autonomous. Loyal to their own research and development priorities, they operate like rival universities, proud of their own traditions and skeptical of the capabilities of other centers.

To keep the centers happy, NASA executives had to tell them how the space station program would be organized—even before it was approved. The key management issues could be boiled down to three questions:

1. Could NASA do the critical systems-engineering work on the space station in-house, or would the agency have to hire contractors to do this work?
2. Who would coordinate the design and construction of the station—a group at NASA headquarters or a program office at what NASA officials called a "lead center"?
3. How many centers would be involved in the actual development of the space station, receiving what NASA officials called "work packages" to help build the facility, and how would the work be divided among them?[1]

In each case, NASA officials answered the question with a solution that maximized the participation of the centers, even when there were other solutions that were better from a purely managerial point of view.

When the Space Station Task Force started its work in 1982, Robert Freitag had created a Program Planning Working Group to address organizational issues. "The job on space station is more difficult than Apollo," Freitag observed, remembering the program upon which NASA's reputation for managerial prowess was based. The Apollo program had what the engineers liked to call "clean interfaces," a reference to the ease with which the various parts of the mission fit together. "The spacecraft sat on top of the rocket," Freitag chuckled, "there was just the launch and then it was gone."[2]

The various components of the space station would snake past each other, making a clean division of work harder to find. That would complicate what NASA officials called the systems-engineering-and-integration function, or SE&I. In addition, the space station would remain in orbit for decades. It had to be organized in such a way as to accommodate growth and accept new technologies. These challenges put a lot of pressure on the Task Force leaders, who were trying to choose the correct organizational form.

Recognizing the importance of the organizational issues, Freitag had decided to chair the working group himself. As his duties as Deputy Director of the Space Station Task Force became more complicated, he turned the chairmanship over to Lee Tilton, who in turn passed it to Jerry Craig as new members joined the group. Craig's group tackled the difficult question of whether to contract out the systems-engineering function, producing a white paper and organizing a management colloquium at NASA's Wallops Island facility held on the last three days of August 1983.

Hiring contractors to start the systems-engineering work had many advantages. This approach, Craig's group wrote, "provides flexibility by allowing activities to be centralized until Center roles and missions are defined." It would give a single group in one place the overall responsibility for integrating the program, especially during its formative stage. That approach, however, cut out the field centers. It would also leave NASA without a corps of skilled employees who understood how the space station fit together. In-house SE&I, Craig's group observed, would provide "a mechanism for enhanced vitality of NASA's in-house technical capability."[3]

There was not much support for contracting out the systems-engineering-and-integration function. "It's going to be a long-term program, and we don't want to put ourselves at the mercy of a contractor for that long a period of time," said Bob Freitag. "There really isn't any single contractor who's got all the capability, and we're not about to build them up with a capability they haven't got." NASA needed that capability itself. Any contractor would also have to coordinate the work of the international partners. "I don't think we or anyone wants a contractor integrating Japan, Canada, Europe, United States."[4]

By relying upon in-house talent at the different field centers to perform the systems-engineering work, NASA officials accentuated the importance of the second question: How would these people be coordinated, and from where? NASA identified the work it did through the identification of levels. Level A—commonly the Washington headquarters—set policy and found the money to run the programs. At level C—commonly the field centers—NASA employees worked with contractors to build the hardware and make it fly. That left level B to coordinate the field centers working on a project and perform the systems-engineering work.

Finding a place to house the level B program office for the space station was a knotty problem. "Back in Apollo," Bob Freitag explained, "the equivalent job was done by Sam Phillips' organization up here in Washington. Sam Phillips' Apollo program office had about eleven or twelve hundred people aboard [most of them contractors], and he integrated the work of Johnson, Marshall, Kennedy, and Goddard." Located in Washington, close to the primary center of power and far enough removed from the field centers, Phillips and his managers could insist upon a program organization that had straightforward lines of accountability.

Freitag gave an example of how the system worked: "When Apollo started out, we looked at the overall system. We said, 'Well, 85,000 pounds is the weight that the Saturn rocket should put into orbit on the way to the moon.'" The Marshall Center and its contractors had to produce a rocket that could deliver 85,000 pounds; the Johnson Center and its contractors had to produce a spacecraft that did not weigh more than 85,000 pounds. Continuing, Freitag said: "They started weighing the spacecraft and the guy comes back and says, 'I can't build that spacecraft for under 120,000 pounds.' And you damn well have to."

"The Apollo guy says, 'I probably can cut the weight from 120 to 110, but it's going to cost me fifty million dollars more than I've got budgeted for.' And the rocket guy says, 'Well, I can raise that from 85 to 95, but I need fifty million dollars more.' "

"So Sam Phillips, sitting on top, has to make a simple decision. Do we modify the rocket to put up 125,000 pounds or do we cut back on the spacecraft to 85,000 pounds or someplace in between?"

"Usually, it's in between," Freitag recalled. "We were spending $10,000 a pound to get weight out of the LEM (the lunar module) by just shaving metal thinner."[5]

A centralized program-integration office worked for the Apollo program because of the "clean interfaces" and the fact that the flights did not last very long. It also worked because Apollo was an approved program with lots of political support, the centerpiece of the agency's mission for the 1960s. The space station program enjoyed none of those advantages.

Largely to resolve the coordination question, NASA Associate Deputy Administrator Philip Culbertson called together all of NASA's center directors and associate administrators, the agency's top career executives. They met at the Langley Research Center on September 22 and 23, 1983. Since 1981, Culbertson had served in the Office of the NASA Administrator as the principal career executive responsible for space station planning. He had assigned the first headquarters employees to work on the space station proposal. Having joined NASA eighteen years earlier during the Apollo era, Culbertson had steadily climbed the career ladder at NASA headquarters, working on a variety of policy, planning, and program-integration tasks.

Culbertson told the career executives that he viewed them as his "executive committee," and that they had been assembled at the Langley Center to provide the advice needed to start the process of organizing the space station program. Jim Beggs and Hans Mark, the top political executives at NASA, planned to join the group at the end of the second day. President Reagan had not yet approved the program, but assuming that he did, the group would have to organize it quickly. Culbertson wanted to identify the major management issues on which members of the group agreed and disagreed and put those issues and the group's recommendations before Beggs and Mark.[6]

The NASA center directors assembled at the Langley Center told Culbertson that they did not want to locate the level B program office in Washington, D.C. Since the end of the Apollo big money days, the centers had grown ever more autonomous and less willing to be directed by anyone. "The centers did not want to work for headquarters," said Frank Hoban, the chief administrative officer on the Task Force.[7]

John Hodge, Director of the Space Station Task Force, wanted to locate the level B organization at a brand new field center, preferably one in a dismal location so that only people who really wanted to work on the station would come. Freitag wanted to locate level B at NASA headquarters or, lacking that, a neutral site like the Langley Research Center, which did not have a major stake in the program.[8]

On the morning of the second day of the Langley Management Colloquium, the center directors left the meeting room and caucused. They emerged from their caucus with a recommendation to put responsibility for the level B management and systems-engineering activities at one "lead center" to which all other centers would report. Johnson Space Center Director Gerald Griffin produced what became known as the "barbed wire" organizational chart to illustrate how the lead-center concept would work. The diagram showed four level C project managers, each developing components of the space station and reporting to their respective center directors. A level B program manager was positioned low in the hierarchy, beneath the center directors. A crooked black line showed that the level B space station program manager would report to his own center director. No thick black lines of authority linked the level A, B, and C space station managers to each other. Instead, barbed wire bound them together. Barbed wire held the level C project managers in line with the level B program manager and headquarters. No lines or wires at all bound the center directors to Washington, or to each other, the first sign that the talk about cooperation might be more rhetoric than substance.[9]

In order for it to work, the lead-center approach would require an exceptional degree of cooperation among people in the field, especially since agency executives wanted to do the systems-engineering work with NASA employees. NASA's engineering talent lay scattered across the country. Center directors would have to let lots of people travel to the level B program center, and those people would have to be willing to go. The center director for whom the level B program manager worked would have to give that person lots of resources with which to do the job. All of the center directors would have to abide by the decisions of the level B program manager, a person who in their eyes was not their superior. As a center employee, the level B program manager would not possess much authority. The center directors insisted that they could compensate for that lack of authority with a spirit of cooperation, but as a center employee, the level B director might turn out to be the tail trying to wag the proverbial dog. Said Sam Phillips, who had run the Apollo program from Washington, D.C., "The lead center concept had been previously tried in NASA over the past 15 or so years with mixed results."[10]

On top of that, there was the problem of international cooperation. Neil Hutchinson, who eventually became the level B program manager, looked upon each international participant "as another field center, each one of them on their own." Hutchinson found that he could not manage the foreign partners from the field. He had to depend upon people at NASA headquarters to conduct country-to-country negotiations on technical details.[11]

The center directors maintained that the plan would work. "Centers can, and do today, 'work for' another center," they maintained. To "assist in the resolution of disputes between level C centers," they proposed the creation of a "Board of Directors" made up of the participating center directors.[12]

Having made the proposal for a lead center, the center directors and the other participants in the Langley Management Colloquium proceeded to vote on which installation it should be. Phil Culbertson told them that their vote was not binding. It was a straw vote, just advisory, yet not something that could be easily ignored. The Johnson Space Center won, overwhelmingly. "Our friendly center directors," Bob Freitag observed, "collectively felt that that was the lesser of two evils," the greater evil being a move to put the level B program officer in Washington.[13]

"The whole purpose was to get the management of the agency together as a group to agree on a management plan," Frank Hoban explained. "There was a strong sense that if they all decided it was the right way to do something, it would be the right way to do it." Hoban praised the process, noting its ability to get the people who ran the field centers to buy into the space station program. "People were afraid not to be there," he said, "because they knew damn good and well we were going to figure out how to manage this doggone program."[14]

William Lucas, the director of the Marshall Space Flight Center, presented the recommendations of the group to NASA Administrator James Beggs, who arrived on the afternoon of the second day, September 23, 1983. Upon hearing the recommendations, Beggs cautioned the participants. We do not have an approved program yet, he told them. The administrative momentum, however, was already under way. The center directors had bought into a management plan that promised little supervision from above in exchange for a promise of cooperation from below.[15]

Beggs approved the management plan on February 15, 1984, three weeks after President Reagan directed NASA to start work on the project. In a letter to Johnson Space Center Director Gerald Griffin, Beggs wrote, "As recommended at our LaRC meeting, I have decided that we shall use the 'Lead Center' management approach. I am hereby designating JSC to serve as the Lead Center for the space station." Among his many responsibilities, Beggs wrote, the level B program director would be in charge of space station systems-engineering-and-integration. On April 9, 1984, while Congress was still deliberating the space station proposal, the director of the Johnson Space Center announced the appointment of Neil Hutchinson as manager of the level B space station program. Hutchinson had spent nearly all of his twenty-two-year professional career at the Johnson Space Center in Houston, working on navigation and flight operations. As previously announced, he would remain at the Johnson Center to organize the space station program.[16]

Hutchinson's appointment tested the degree to which the center directors would really cooperate. Hutchinson had to integrate the work of the level C project managers, who, working with their contractors, had to design and build the various components of the space station. But before Hutchinson could integrate their work, NASA officials had to answer the third key management question: How many centers would receive the so-called "work packages" and how would the packages be divided up?

"Every inch is worth about $10 million bucks," Frank Hoban said as he watched the center directors fight over the shape of the work packages. "They were fighting for every inch of that thing." Jim Beggs told Philip Culbertson to oversee the work-package assignments. Culbertson agonized over the process. "The agony was how to divide them up," he said.[17]

Culbertson wanted to divide the space station program up into three work packages. The Goddard Space Flight Center in Greenbelt, Maryland, would handle the free-flying platforms, complementing that installation's traditional role as a science center. The platforms were conveniently detached from the main part of the space station, and that in itself reduced the problem of coordination. As for the main part of the station, Culbertson hoped to divide it up between the Johnson Center in Houston and the Marshall Space Flight Center in Huntsville, Alabama, the two centers that had been primarily responsible for the Apollo program.

Culbertson wanted to keep the division of work between the Marshall and Johnson centers as simple as possible. He wanted to keep the division clean so that each center could do its work as independently as possible. He wanted to balance the assignments so that neither center would get more work than the other. And he wanted to avoid giving the Marshall Center a level C engineering responsibility for any task that would tend to integrate the space station, such as the data-management system. Neil Hutchinson, the program manager at the Johnson Space Center, was having enough trouble integrating the components of the space station as it was. If the engineers at the Marshall Center were given responsibility for a task that permeated the entire station, that would put them in a position to do Hutchinson's work.[18]

Culbertson's hope for a simple, three-way division of responsibility quickly turned into a four-way contest for work. The director of NASA's Lewis Research Center in Cleveland, Ohio, Andrew Stofan, announced that he wanted his center to develop the station's electric power generating plant. Culbertson thought that a fourth field center would complicate the process too much. Stofan, however, had lots of support, both inside NASA and on Capitol Hill.

As he studied the possible configurations of the space station, Culbertson discovered how hard it was to draw clean lines of separation. Even though the solar collectors for the power system could be placed on booms far away from the live-in core, the electric lines and equipment permeated the whole facility. Both the Johnson Center and the Marshall Center wanted to develop the modules out of which the main space station would be built. Officials at the Johnson Center wanted to equip the inside of the crew module, since they were responsible for training the astronauts. Officials at the Marshall Center wanted responsibility for the outside of the crew and laboratory modules, since they had built Skylab. Try as he could, Culbertson could not make the two centers agree on where the inside ended and the outside began. Each center tried to define its own area of influence as deeply as it could. "I could not get everybody to agree to a single solution," Culbertson confessed. "There were some bitter battles about where that dividing line was."[19]

Before he assigned the work packages, Culbertson had to distribute responsibility for the advanced-development program. Growing out of the earlier work on space station technology, the advanced-development program challenged the field centers to create what NASA engineers called "test beds" to examine new techniques that might be used on the space station. The center directors were very interested in the test-bed concept because it meant both money and technical responsibility for their centers.[20]

The leaders of the Task Force decided to hold a competition for the test-bed assignments, asking the centers to submit proposals as if they were contractors making bids. James Romero, a member of the Task Force, organized the competition. An independent committee evaluated the proposals, and Culbertson made the selections, focusing primarily on the technical strength of each center rather than on the more elaborate criteria he had developed to balance the work-package assignments. He announced the results on February 29, 1984. Hoping to preserve flexibility, Culbertson awarded each of the seven test-bed assignments to a multi-center team, a tactic that would test the degree of cooperation the center directors had promised to deliver.[21]

The center directors, however, were more interested in the designation of the lead centers for the advanced-development activities than in the naming of multi-center teams. They viewed the advanced-development assignments as the first cut for the more sumptuous work-package awards. "The mistake I made as I went through that process," Culbertson later explained, "was that I didn't fully appreciate that in making those assignments I was really limiting the flexibility which I would later have to make work-package assignments." What Culbertson and the Task Force hoped would provide a basis for multi-center cooperation turned into what one center director called an unbelievable battle for turf, with each center fighting to hold on to what it had already won.[22] One key member of the Space Station Task Force got so mad at the rigidity being built into the space station program that he quit the Task Force and left town. The center directors were locking NASA into a premature division of responsibilities before they had even developed a preliminary design for the space station. "The goddamn centers won't sit there and work for another center," a member of the Task Force complained.[23]

The lead-center assignments for the advanced-development program went to the Johnson and Marshall centers, three each. Obviously missing was the designation of a lead center for the electric power program. Andrew Stofan, the director of the Lewis Research Center, expected to get this assignment. He, however, was locked in a struggle with the director of the Marshall Space Flight Center for this responsibility. Culbertson announced that this lead-center assignment would be "designated later."[24]

Facing pressures from Task Force leaders and the center directors to broaden the base of agency participation, Culbertson finally agreed to a four-way split for the work-package assignments. While noting that John Hodge "would have favored having every center involved in some way," Culbertson observed that "it simmered down to a decision to have four." Even so, Culbertson and Hutchinson could not get the center directors to agree on an exact division of responsibilities. Culbertson made the final divisions himself.[25]

On April 6, 1984, NASA created an interim Space Station Program Office, a level A headquarters group to set overall program policy and get the resources to build the station. Jim Beggs appointed Culbertson to head the interim office and made John Hodge his deputy. Robert Freitag, Terence Finn, Daniel Herman, and other members of the Task Force took key positions with the interim office. Beggs would have created a permanent space station office at NASA headquarters, but Congress still had not approved the program.[26]

The previous month, the Task Force finished its last piece of work, publishing the final edition of the six-volume "Space Station Program Description Document," the so-called Yellow Books. For the first time, Task Force members were able to get NASA to put yellow covers on them, a small management victory in what had otherwise turned out to be a troublesome beginning.

23
Congress II

Without much debate, Congress approved legislation authorizing NASA to start work on the space station. Authorization of the space station, however, only started the legislative battle. Congress could kill what it had authorized by not appropriating the money to fund it. If opponents of NASA's space station wanted to stop the program, they would make their stand in the House appropriations subcommittee where the funding bill began. In that subcommittee, both the chairman and the ranking minority member opposed NASA's plans.

Terry Finn looked at the names of the nine members of the House appropriations subcommittee and counted. As a congressional staff aide and later as NASA's Director of Legislative Affairs, he had gained skill at the art of counting votes. Now, as a member of the rapidly disappearing Space Station Task Force, he could not find five people on the nine-person subcommittee ready to approve funds for a permanently occupied space station.

Congressman Edward Boland, the Massachusetts Democrat who headed the appropriations subcommittee, wanted to fund nothing more than a "man-tended" space station. Instead of a multi-modular, 75-kilowatt space base, Boland would allow NASA to launch a single, multi-purpose module equipped as a research laboratory. Astronauts would work, but not live, in the laboratory module. For living quarters, they would rely on the shuttle orbiter, docked to the research lab. At the conclusion of each twenty- to thirty-day mission, the orbiter crew would separate from the laboratory and return to earth, leaving the module unoccupied until the next crew arrived. As later defined by NASA, the laboratory would have its own power supply—37.5 kilowatts of electricity generated by solar arrays—and a life-support system that would work in conjunction with the environmental-control equipment in the orbiter.[1] Boland believed that NASA could complete 80 percent of the 107 missions proposed for the permanently manned facility by building two unmanned platforms and the man-tended laboratory. This could be done, Boland insisted, "at only 15 or 20 percent of the cost" of a permanent station.[2]

Congressman Bill Green, the top-ranking Republican on Boland's subcommittee, agreed. Even though he represented President Reagan's party on the nine-member subcommittee, Green did not support the administration proposal. Elected from a liberal congressional district in midtown New York, Green spoke for the space science intelligentsia opposed to the permanently occupied facility. A manned space station, Green said, would "put the squeeze on other vital scientific programs." Congress should not, he continued, "permit one program to jeopardize other less dramatic but valuable scientific exploration." A manned space station might be possible "at some unspecified but later date," he observed, but for now NASA should confine itself to unmanned platforms. Opponents of the permanently occupied space station were generally willing to admit that the United States might build such a facility someday, so long as the government did not actually commit funds to the project now.[3]

A dozen years earlier, the position of the chairman and the senior minority member would have prevailed. Until the mid-1970s, civilian space policy had been set by agency executives working with the President's staff and a few key congressional committee chairmen. In 1975, however, legislators lacking in seniority had staged an uprising in the House Democratic caucus and forced the old guard to accept the right of the majority to elect committee heads. Combined with the continuing erosion of party discipline, these reforms had "democratized" the House, significantly reducing the power of a few well-placed lawmakers to set public policy. NASA officials now had to deal with all of the members, not just a few committee chiefs. This forced NASA officials to broaden their base of support. In the case of the space station, it also created an opportunity. Supporters of the permanently occupied space station could override the wishes of Boland and Green, if only NASA officials could win over the other members on the committee.

Finn spoke to Hans Mark, Deputy Administrator of NASA, who was hearing much the same message from his contacts on Capitol Hill. "It was clear," Mark observed, "that the game wasn't over."[4]

To win approval for the space station, NASA officials had to meet with individual members and their staffs and educate them on the future of the space program. NASA could not count on White House officials to do this. "They rely on the executive agencies," Mark explained. Nor could NASA executives wholly depend upon their own offices of legislative and budgetary affairs, set up to answer congressional queries and handle routine legislation. This was the first time NASA officials had been obliged to present a major space flight initiative since the new congressional reforms, and they needed a special effort. Mark took charge of a legislative-strategy group. Philip Culbertson and Norm Terrell represented the Office of the Administrator; Terry Finn and John Hodge led a delegation from the Task Force. Working with people from NASA's Office of Legislative Affairs and the agency's budget office, the group identified the legislators NASA would have to contact and the best way to explain the program to them.

NASA was not the only agency attempting to educate lawmakers on the space station issue. Congress had asked its own Office of Technology Assessment, in the words of the members of the House Committee on Science and Technology, to "undertake an independent, rigorous, balanced study of the need for a space station." In a detailed, 129-page staff report, issued four months *after* the congressional debate ended, OTA officials explained why Congress should reject NASA's proposal. Based on their research, the OTA staff concluded, "the present NASA 'space station' concept is not likely to result in the facility most appropriate for advancing U.S. interests into the second quarter-century of the Space Age."[5]

Three months *before* the crucial House vote, the project director of the OTA staff study was called to testify before the committees studying the proposal. To vote against the space station, the lawmakers needed a good rationale, and the analysts in the Office of Technology Assessment seemed to be in a good position to provide it. Disorganization, however, prevailed.

"Let me underscore at the outset," project director Thomas Rogers said as Congress considered the program, "that OTA does not take a position on the desirability of any space station." Members of Congress were shocked. "Should this not be one of the areas where you ought to have some real advice?" Senate Budget Committee Chairman Pete Domenici asked. "It is impossible," Rogers stated, "to judge objectively whether or not NASA's currently favored infrastructure elements are truly appropriate and worth their substantial public cost."[6]

Slade Gorton, Chairman of the Senate Subcommittee on Science, Technology, and Space and a strong supporter of the permanently manned space station, put the question directly to Rogers. Should Congress, Gorton asked, "delay for another year or two the initial authorization?"

"Oh, no," Rogers replied. "The future is always the enemy of the present, Mr. Chairman. And we do not suggest for a moment, we would not want to be thought of as suggesting that."[7]

A second helping hand came from Thomas M. Donahue, Chairman of the Space Science Board of the influential National Academy of Sciences. In 1983, Donahue (a professor of atmospheric science) had written to NASA Administrator James Beggs to inform him that "the Board sees no scientific need for this space station during the next twenty years."[8] After a series of conversations with Beggs, Donahue communicated a somewhat revised position to appropriations subcommittee chairman Edward Boland.

"I cannot emphasize too strongly," Donahue told Boland, "that it is *not* correct to characterize the Space Science Board as being 'opposed to the Space Station.' In fact I am persuaded that there are many ways in which the Space Station will enable crucially important space science investigations of the future." Inside NASA, he wrote, "there is no disposition to compromise science missions by forcing them on to unsuitable platforms and no disposition apparent at this time to torque priorities." Members of the Space Science Board had "received firm assurances from Mr. Beggs and his associates that NASA understands the concerns we have articulated."

"I hope that you agree with me and with the administrator of NASA," Donahue concluded in his two-page letter, "that space science can indeed prosper during the Space Station era and not only live with the Space Station but make effective use of it. Obviously this cannot happen without your cooperation."[9]

Opposition to the space station was not well organized. The science community appeared to be split on the space station issue, exactly the impression that NASA wanted to create. NASA, on the other hand, was organized. Any large government bureaucracy has the resources to mount an effective legislative campaign, assuming it has the skill to do so. By now it was clear to Mark and Finn and the other members of the legislative-strategy group that NASA would have to do just that.

Administrative agencies work the Hill by going out and getting help, always a delicate task. Civil servants are prohibited by law from pressuring members of the public to influence Congress on behalf of agency programs. NASA, however, had already enlisted the cooperation of aerospace industries by contracting out the mission analysis studies, a permissible act. Representatives of the aerospace industries in turn contacted members of Congress to explain the results of their studies.[10]

John Yardley, President of McDonnell Douglas Astronautics Company, praised the potential for the manufacture of pharmaceutical products in space, an enterprise the company had been testing on the space shuttle. "Only manned presence allowed the successful demonstration of continuous flow electrophoresis to process high-density material," Yardley wrote to Congressman Boland. "Potential commercial investors in space and the manned space industry are looking for a strong signal from Congress of their support for an aggressive, permanently manned space program."[11]

Paul Burnsky, President of the Metal Trades Department of the American Federation of Labor and Congress of Industrial Organizations (AFL-

CIO), told the members of the House appropriations subcommittee that "labor is proud of the significant contributions which our skilled craftsmen have made in our space programs. We look forward to the decade of the 1990's and the fulfillment of the human and scientific advances for all mankind that will result from the successful operation of the U.S. manned space station."[12]

NASA officials met with the staff of the Congressional Space Caucus, an informal group of 166 representatives who joined together in 1981 to help revitalize America's space program. Congressman Daniel Akaka, a Democrat and co-chair of the space caucus, sent out a detailed, four-page letter urging the members of the House appropriations subcommittee to fund the manned space station "at its full start-up level of $150 million for fiscal year 1985." The letter sent a not-too-subtle message that the members of the appropriations subcommittee could expect a floor fight if they tried to replace the fully manned space station with a man-tended platform.[13]

As the Boland subcommittee prepared to vote, NASA's strategy group continued to schedule meetings with the legislators and their staffs. "During the next two or three weeks, Jim Beggs and I, between us, met at least once with each member of the subcommittee to explain to them the details of the NASA program," reported Hans Mark. Staff-to-staff contacts also were important. Not only did members of the legislative-strategy group have better access to young staff members than to the lawmakers themselves, but the lawmakers relied more on their own staffs to help them make decisions as the power of party leaders and committee chairs diminished.[14]

Jim Beggs summarized NASA's objections to the "man-tended" space station in a letter to Congressman Boland as the key committee vote neared. The effort needed to develop such a facility, Beggs wrote, would be "costly and dead-ended." People were needed in space for the same reason they were needed in technical facilities on the earth—"to permit an effective response to unanticipated findings," Beggs explained, "and to maintain and repair those things which are not amenable to pre-programming." Although the space station would be highly automated, that automation would demand "detailed manned supervision."[15]

In the Cabinet Room, Beggs had deflected concern over the cost of the station by impressing President Reagan with the commercial potential of space—and the role a space station would play in preserving America's leadership in technology. Democrats in Congress, even those who were inclined to support NASA's vision of human flight, were not as entranced by the vision of astronauts toiling in microgravity space factories. The lawmakers seemed much more interested in the implications for international cooperation on a multi-national space station. "One of the arguments that we used," Mark explained, "was the fact that the space station program would be done on an international basis."[16]

"I am very interested in the discussion about the U.S./European/Japanese collaboration," Democratic Congresswoman Lindy Boggs told Jim Beggs. One of the swing votes on the House appropriations subcommittee, Boggs asked Beggs about the free-flying space station module being developed for the European Space Agency. "Do you think that Columbus will, indeed, be a part of the U.S. space station?" she inquired. Boggs worried whether NASA's schedule for constructing the space station would be sufficiently fast-paced to keep the Europeans on board. If the United States faltered, Europe might attempt to launch the Italian-built Columbus module in 1992, the 500th anniversary of the first voyage of Christopher Columbus. "Apparently we are going to depend upon another Genoese to make discoveries for America," she warned.[17]

"What is their interest?" Representative Bob Traxler asked Jim Beggs. "What are they prepared to do? What did you ask them to do?" Traxler was the second-ranking Democrat on the House appropriations subcommittee. A representative from the Bay City area of north central Michigan, he too expressed interest in the possibilities for international collaboration with the Europeans and Japanese. "Do they have the same degree of interest in an unmanned station as they may have in a manned station?"

Beggs used the opportunity to encourage Traxler's support. While admitting that the Japanese were "extraordinarily interested in robotics and automation," Beggs explained that the Europeans had discovered "the importance of the man in the loop." The flights of European astronauts aboard the American space shuttle had been very well publicized back home. "They like the idea of being part of a program which in contrast to most of the R&D, which gets little or no publicity except in scientific journals, is front page news." If you proposed an unmanned space station to them, Beggs predicted, "I don't think you would get as much interest as you would in the case of a manned station."[18]

The third Democratic vote, Congressman Lewis Stokes, was interested in a domestic issue. Stokes represented the city of Cleveland, Ohio, where he had served as mayor before his election to the U.S. House of Representatives. NASA's Lewis Research Center, with its 2,800 employees, sat on the west side of the Cleveland municipal airport. The director of the Lewis Research Center wanted the work package for the electric power generating system on the space station. "Of the seven center assignments," Stokes said, referring to NASA's distribution of the advanced-development responsibilities, "only one function out of the seven did not have a designation of a lead center and that area was the electric power system area." Stokes wanted to know why. "In that area, you set up what is known as an intercenter team." When did NASA plan to name a lead center? "Are you in the process of developing some special criteria for the lead center?" Would economic considerations be used? What about management capability?[19] NASA officials eventually awarded responsibility for the electric-power-generating system to the Lewis Center.[20]

The legislative-strategy group continued to meet with undecided subcommittee members and their staffs. "Let's get to the million dollar question, or perhaps the billions of dollars question," Congressman Traxler urged, searching for an approach that might bridge the two positions on the committee. "Congress likes to keep its options open. We have some variables that those of us on this committee are not accountable or responsible for. Congress may make decisions with respect to the Administration concerning taxes, concerning budget cuts and we may be forced with certain decisions that perhaps the members of this committee would not want to make otherwise," he said, echoing Chairman Boland's concerns.

Traxler directed the question to Philip Culbertson, who was sitting at the witness table with Jim Beggs and Hans Mark. Culbertson, who had also been meeting with the members and their staffs, sensed that the committee position was shifting. "If we give you $150 million," Traxler asked, "are you locked into only a manned program?"

"No," said Culbertson. "The things that we will be doing with that $150 million will include most of what we would do if we were going to an unmanned program."

Traxler asked again to make sure. "So the $150 million appropriation does not mean that the program is absolutely locked in to a manned station?"

"No," Culbertson replied. "It means we are trying to understand better what the manned system would be."

"Thank you," Traxler said.[21]

The members of Boland's subcommittee met on May 14, 1984, to mark up the appropriations bill. Boland and Green favored the man-tended option. They had support on the Senate side of the Capitol, where Appropriations Committee staff members had prepared language prohibiting NASA from spending money on any manned components of the space station until agency officials had studied a fully automated facility. Boland and Green were prepared to lead the fight against a permanently occupied space station, if a majority of the members of their House Appropriations subcommittee would let them.[22]

They would not. A majority of the members urged Boland to drop his opposition to NASA's proposal, which Boland refused to do. If Boland and Green offered a motion to substitute their man-tended facility for a permanently occupied space station, they would lose that vote, so they did not call for one. Within NASA, this was viewed as a major victory. Through their efforts on Capitol Hill, NASA officials had won enough votes to defeat any motion that would force the agency to build a space station run by machines or occasional visitors.

Boland, however, still had enough influence to prevent his committee members from endorsing NASA's vision of a permanently occupied facility. In its place, the subcommittee reported out a bill that allowed NASA to start the phase B definition and preliminary design studies on "a space station." As part of that appropriation, the subcommittee required NASA to spend "not less than $15 million for complementary space station studies to define an alternative concept employing an initial 'man-tended' capability rather than a 'permanently manned' capability."

"The committee believes," Boland inserted in the report to the House of Representatives, "that if future deficits do not permit the full development of the station—it is essential that the permanently manned element not be the principal or sole survivor of budget retrenchments." Given that possibility, "it is important for NASA to define an option which 'phases in' the permanently manned feature of the station." Under the "phased-in" approach, NASA would build the space station by launching the laboratory modules first, adding on the habitat modules as assembly progressed. "We are," Boland warned during the floor debate, "as sure as day follows night, going to face some kind of budget retrenchment in the future. When that retrenchment comes, what we do not want to see is the permanently manned element of the space station as the sole survivor." Boland asked what would happen if NASA launched the habitat modules first and ran out of money for the laboratory modules with the space station half-built. "Are we going to end up with a hotel in space? That's what worries me."[23]

By requiring the man-tended study and the phased-in approach, Boland left open the possibility that events would force NASA to do in the future what he and Congressman Green lacked the votes to make the agency do in 1984. NASA officials clearly wanted to close off this possibility. "NASA is concerned," James Beggs wrote to Senator Jake Garn, "about the [House] Committee's direction to study an additional space station option. A space station that is not manned full-time from the start would be far less capable than a permanently manned space station; such a man-tended space station would not ensure leadership in space for the United States during the 1990's; and man-tended space station capability will be more costly to develop, if the plan is to eventually obtain a permanently manned capability."[24]

An internal NASA memorandum was even more pointed. "The President's directive to NASA in the State of the Union did not envision the study of two space station options but called for the development—within a decade—of a permanently manned space station." NASA officials argued that such a proposal "would send the wrong message to Europe and Japan who are about to commit to and are interested only in a U.S. station that is permanently manned." The Marshall Space Flight Center, they added, had already "spent over $8 million in 1979–83 looking at space platforms with manned capabilities, essentially what the House has directed us to study." Such a facility, agency officials concluded, "is not the appropriate space station to develop."[25]

When the appropriations bill reached the floor of the House of Representatives on May 30, 1984, Congressman Robert Walker moved to strike the directive for a "man-tended" study. "I think we need a firm commitment to a permanently manned station in outer space," he said. "I do not think we ought to be negotiating on that point." A Pennsylvania Republican and member of the House space science and applications subcommittee, Walker forced the manned versus man-tended issue. "It is really a question of the definition of this space station," he explained, "whether or not we are really going out to design a permanently manned space station or whether or not we are going to be willing to settle for a man-tended space station as a possible alternative."

Congressman Boland explained his position. "All we are doing is telling NASA, 'Look, as part of the RFP [Request for Proposal] for a permanently manned station also define what the station would look like with a man-tended approach.'" NASA had already agreed to spend two to three years studying the requirements of the space station and preparing a preliminary design. The agency would not be ready to give out the industry contracts to start building the components of the space station for three years. "When we come to the 1987 funding request for actual development of the space station," Boland advised, "we will at least have a chance to look at the two options."

Congressman Green agreed. "The choice is very simple," he said. "We shall be told some alternative ways to proceed with a space station. We shall then have some choices as to the sequences by which to proceed."

"I concur with the language that the Committee on Appropriations has included in the bill," Congressman Don Fuqua announced. As chairman of the Committee on Science and Technology, he had praised the permanently manned space station. "I think, though, that it is prudent to consider all the alternatives while we are in the study phase."

"I wish to echo the remarks of the chairman of the full committee," said Congressman Harold Volkmer. As chairman of the House Space Science and Applications subcommittee, Volkmer was one of NASA's best allies on Capitol Hill. "I think that it is very noteworthy to have it [the man-tended study] remain in there so that we can have these studies during the phase B period."[26]

Had NASA officials been able to make the decision themselves, they would have scrapped the man-tended study. To properly conduct a long-term technology program, they felt they needed to pursue the fully occupied space station from the start. The lawmakers, however, wanted to proceed more cautiously, step by step, and NASA had given them the opportunity to do so by proposing to build a space station that would grow incrementally and by promising to study the concept for two to three more years before starting to build it.

Congressman Walker asked the House of Representatives to make a commitment to a permanently occupied space station then, in 1984. "I think that that commitment is absolutely essential," he said. By a voice vote, his motion was defeated.[27]

As the appropriations bill wound its way through the Senate and into conference committee, NASA was able to tone down the language mandating the study of a man-tended station somewhat. Boland's bill required NASA to spend exactly $15 million on the study. Meeting with the staff directors of the two subcommittees, Philip Culbertson worked out compromise language that dropped the reference to both the specific dollar amount and the phrase *man-tended,* substituting instead the requirement that NASA "conduct a study of an option which 'phases in' the permanently manned features of the station."[28]

The verbal gymnastics had little effect. When NASA submitted the required study two years later, it carried the title *Space Station: A Man-tended Approach.* The lawmakers also directed NASA to complete a second study on space station automation and robotics and spend no less than 10 percent of the cost of the facility on such systems.[29]

On July 16, 1984, as part of the bill approving all of NASA's FY 1985 activities, President Reagan signed the legislation authorizing what NASA officials believed would be the first step toward a permanently occupied space station. Two days later Reagan signed the appropriations bill authorizing the expenditure of $155.5 million to start the project. The following month the White House issued a new National Space Strategy, the first major revision in White House policy since President Reagan had made his July 4, 1982, commitment to "a more permanent presence in space." In spite of congressional misgivings, the White House document affirmed Reagan's decision to let NASA build "a permanently manned space station within a decade" and to seek international cooperation in its "development and utilization." In a sign that the incremental epoch of space policy might be drawing to a close, the document also affirmed the decision of Congress and the President to appoint a National Commission on Space to identify long-range goals.[30]

On August 1, 1984, NASA Administrator James Beggs established a permanent Office of Space Station at NASA headquarters. He appointed Philip Culbertson to head it as Associate Administrator for Space Station and confirmed John Hodge as Culbertson's deputy. Daniel Herman became Director of Engineering, Robert Freitag was named Director for Space Station Policy, and Terence Finn became Freitag's deputy. In the Nova Building, part of an inconspicuous group of two-story white-brick office buildings adjacent to the Johnson Space Center, a special team of NASA engineers and scientists began to draw pictures of what the space station might really look like.[31]

Afterword

Politics, Bureaucracy, and Public Policy

In spite of the lack of political consensus about the future of the U.S. space program, and in spite of the large federal budget deficit, NASA officials were able to get a space station approved in 1984. Official approval, however, did not remove the space station issue from the political arena. The space station continued to face political difficulties as NASA officials worked to design and build it.

The difficulties were frequent and varied. Station critics persisted in their opposition, offering alternatives that varied from "man-tended" laboratories to outright denial. Money woes wore down NASA engineers as Congress appropriated less than 40 percent of the funds President Reagan had originally approved for the facility, stretching out the program and forcing design changes. On January 28, 1986, the space shuttle Challenger blew up, calling into question NASA's readiness to take the "next logical step." In 1987, White House officials announced that the cost of the program had soared to $14.5 billion, an 81 percent increase over NASA's original $8-billion estimate prior to the construction of any major part of the facility. Investigations of the Challenger accident exposed deficiencies in the management of the space station program, forcing NASA officials to scrap the fragile "lead center" approach they had set up in 1984.

The desire to build the space station did not magically turn into a steady technology-development program with the appropriation of the first $150 million in 1984. The program remained very much a policy problem, open to amendment and reconsideration. Differences over space policy were not resolved by the space station decision; they were merely postponed.

Many of the difficulties that NASA officials faced while trying to keep the initiative on track arose from events external to the program itself: most particularly, problems created by the persistence of the budget deficit. Just as many difficulties, however, could be traced to the approach agency officials took in trying to get the program approved.

In seeking approval for the space station, NASA officials had to answer many important questions. Why did the United States need a space station? What would a space station do? What would it look like? And what would it cost? The answers to these questions were shaped by the agency's political situation and by its past experience in getting other programs such as the space shuttle approved. Inexorably, NASA officials were drawn toward incremental answers to important questions. In the short run, this made program approval easier. In the long run, it allowed politicians to tinker with the program after it was approved.

Curiously, NASA officials persisted in the use of incremental techniques at a time when the biggest spenders in the federal government had moved on to more sophisticated funding methods. By the mid-1980s, most big-spending federal agencies had escaped the insecurity of incrementalism as their advocates locked long-term commitments into law, creating legal obligations that required Congress to appropriate their funds. Agency advocates did this through entitlements (legal obligations that required the payment of benefits to any eligible person), earmarked revenues (taxes committed to a specific purpose), and credit obligations (the creation of large financial obligations in the future without big cash outlays in the present). NASA's projects, on the other hand, languished among a shrinking number of "discretionary programs" whose budgets could be adjusted by the President and Congress each year. By the mid-1980s, less than 20 percent of all federal appropriations were decided through the "old politics" of discretionary budgeting. Though they sought to remain on the leading edge of science and technology, NASA officials found themselves in a backwater of budgetary review. NASA officials also discovered, as they moved funding proposals for the space station through the budget review process, that many participants no longer observed the rules that guided the old incrementalism. The old rule of reciprocity, in which different parties tended not to interfere in each other's budget requests, gave way to a new level of conflict over scarce resources. The space station was the first major program in which the Department of Defense worked to stop a NASA initiative. NASA officials had always had to contend with grumbling from White House budget analysts and the science community. The opposition of the Defense Department, however, created a level of dissensus that had not existed before. It seemed that NASA officials learned how to play the game of budgetary politics just as the rules were changed.[1]

In the 1960s, during the first full decade of space exploration, the voyage to the moon had given the U.S. space program a purpose and a reasonably solid political commitment. In the two decades that followed the first lunar landing, the U.S. space program settled into the swamp of incremental politics. NASA and its allies continued to push for an aggressive space program, but always against the inertia created by the absence of long-range objectives for U.S. space exploration. Through experience, NASA officials learned to adopt incremental approaches to policy problems. In the space station decision, this occurred in the following ways.

Advocate Initiatives, Not Options. NASA officials placed their plans for a permanently occupied space station on the national political agenda as a single initiative, strongly advocated, not in the more politically neutral manner as a choice among options. Like other incremental decisions, this emerged from the common experience of NASA officials rather than as a result of extensive strategizing.

The terms under which the Apollo and space shuttle programs had been approved were important parts of NASA's institutional memory. By 1981, the twenty-year-old decision to go to the moon had taken on mythical qualities within the agency. The Apollo decision was viewed widely as a "comprehensive" decision, not an incremental one. President John F. Kennedy selected the best alternative from a list of competing options, as the process of comprehensive decision making requires. NASA officials gave technical advice on the options, a posture made easier by the fact that Kennedy's preference for a dramatic goal coincided with NASA's. Once the decision was made, President Kennedy gave NASA officials enormous technical latitude in which to plan the mission and he gave them the resources necessary to carry it out. The overall goal was unambiguous; the criteria for success was clear-cut. It was the first—and the last—unequivocal commitment for a major space program that NASA would receive in its first thirty years.

Experts on space policy, NASA employees among them, yearned for a return to the days of Apollo. As a research-and-development agency, NASA seemed to work best when guided by an Apollo-type objective. The conditions under which Kennedy approved the Apollo program allowed NASA officials to give the President realistic advice about the technical difficulties of a voyage to the moon and the options to it. NASA officials were able to give the President a realistic estimate of the costs.

Why, then, did NASA officials not present the space station to President Reagan in a similar way? The answer, in short, was that they had learned not to do so. In 1969, proud of having met the goal of taking humans to the moon, NASA officials trotted out their long-range plan for the exploration of space. The plan contained nineteen distinct initiatives: a space station, a space shuttle, a lunar base, a large orbiting observatory, an outer-planet tour, a survey of the asteroid belt, an expedition to Mars, and more.[2] The results, for NASA, were disastrous. One of the surest ways to kill a long-range plan is to smoke it out before its advocates have lined up the necessary support. President Richard Nixon, to whom the plan was presented, rejected it. As a consolation, he encouraged NASA officials to pick one program from the list and advocate it. Faced with a choice between the Space Transportation System and a space station, NASA officials selected the former. They also decided to pursue the remaining elements in their unapproved long-range plan incrementally, one by one, waiting to advance the space station until the shuttle became operational.

This incremental approach to space policy seemed more promising given the situation in which NASA officials found themselves. For one thing, presidents from Nixon to Reagan tended to expect it. While professing in public their desire for a subservient bureaucracy, they rewarded agency heads who acted as advocates for limited new starts. NASA administrators could win approval for new starts in the absence of any consensus about long-range goals by advocating single initiatives rather than many options.

Without goals, one cannot compare options. In reviewing his options, President Kennedy sought to realize his goal of a space program that promised "dramatic results in which we could win."[3] No such overarching philosophy guided the U.S. space program in the second and third decades of space exploration. This experience taught NASA executives to eschew options, no matter how much they yearned for the days of Apollo.

Build a Clientele. Politicians commonly expect civil servants to produce a clientele that is ready to support agency programs and help agency heads fight for funds. The only good substitute for a potent clientele is a strong objective to which the public at large subscribes. Unable to secure a long-range objective as justification for their activities, NASA officials concentrated on building a clientele. "One of the things we did," said Task Force leader Daniel Herman, "and in a way it was calculated and I think it's a major reason why the space station has survived, is very carefully build up constituencies to sustain it."[4]

NASA officials sought to include within their clientele entrepreneurs who were entranced by the potential for commerce in space; technicians who wanted a space station from which to repair satellites and launch objects to other orbits; scientists who wanted to study the earth and heavens; military officers and arms-control analysts who needed platforms for orbital reconnaissance; international partners who wanted to share in the benefits of space technology; and space buffs who wanted to explore and eventually settle the solar system.

NASA officials needed a very large clientele. They needed a large clientele to show broad-based support for the space station and to deflect the arguments of opponents who claimed that nobody but NASA wanted a space station. A large clientele, however, would not fit on a single space station. The international partners did not want to be on a space station with the U.S. military. Space scientists did not want to share a space station with astronauts repairing satellites. People studying the earth wanted to be in polar orbit; people studying the heavens preferred an orbit near the equator. Space explorers wanted a station with artificial gravity, space manufacturers wanted one with no noticeable gravity at all.

Not only did people disagree over long-range goals for space, but the ones who supported a space station disagreed about what kind of facility to build. A long-range exploration objective would have resolved those disagreements, much as the Apollo decision had resolved the status of a space station twenty-three years earlier. Without any such objective, NASA officials had to unify their supporters in a more creative way. For that reason alone, they had to adopt an incremental rather than a comprehensive approach toward the initiative in order to get it approved.

An incremental policy, by its nature, does not require participants to agree upon long-range goals. A comprehensive decision does. In a comprehensive decision, such as the decision to go to the moon, policy makers compare the merits of various alternatives. To compare alternatives, policy makers must agree upon the long-term objectives against which the alternatives are judged. The makers of an incremental policy are not so constrained. They merely need to review a limited change in the status quo on the basis of what has already been done. A "good" incremental decision is one that the involved parties agree is good, without any reference to long-term goals. Each party to an incremental decision may retain its own particular set of goals so long as each is led to believe that its goals might be realized upon the completion of the program.[5] As one of the leaders of NASA's Task Force observed, the space station program was carefully crafted to be whatever the users wanted it to be, a consequence of the recognition that a successful program would have to appeal to a diverse collection of users. To minimize the potential for conflict among users with different goals, the leaders of the Task Force made two very important moves.

Build Manned and Unmanned Elements. Task Force leaders understood that the interests of a large group of users could not be accommodated on a single, large space station. The users understood this too. As one way of resolving this problem, Task Force leaders conceptualized a space station program consisting of many parts. They described it in their planning guidelines as "manned and unmanned elements," a concept advanced by Task Force leader Daniel Herman.[6]

Historically, people had viewed a space station as a single thing—most commonly, a large wheel rotating in space. Task Force members included within their definition of the space station program not only the base occupied by people but two automated platforms as well. The main space station with its laboratories, experiment mounts, and servicing bays would travel in a 28.5° orbit some 250 nautical miles above the surface of the earth. So would one of the unmanned platforms. The other unmanned platform would circle the poles. The automated platforms helped meet the needs of the science community, whose interests could not be contained on a single, live-in core. Even so, Herman observed, "we never won the hearts and minds totally of the scientific community."[7]

Including the automated platforms within the overall program did not eliminate all of the potentially conflicting requirements. On the main space station, for example, the requirements of space transportation conflicted with the requirements of space manufacturing. The docking and fueling of spacecraft in transit would cause the station to bump and move around, spoiling the near zero-gravity conditions prized by space manufacturers. Any attempt to design a fully developed space station that met both of these requirements would ultimately require "branching," the further distribution of space station architecture through the construction of two or more main facilities. In the initial stages of development, a single space station would suffice. As uses evolved, branching would surely occur.[8]

Such an admission during the approval process would have severely damaged NASA's chances of getting the program approved. The last thing either opponents or supporters of the program wanted to hear was that NASA wanted to build two main stations rather than one.

As NASA officials designed the space station, inconsistencies such as these would have to be resolved. Given the absence of any well accepted long-term goals, NASA officials correctly perceived that the approval process would be a poor place to resolve them. Premature discussion of design details would simply divide NASA's supporters, making program approval unlikely. In what was surely their most important decision, the leaders of the Space Station Task Force devised an approval strategy that avoided the potentially divisive design issue.

Emphasize Missions, Not Configurations. In one of their most important moves after starting up the Task Force, John Hodge and the other leaders of the group decided to identify the missions to be performed on the orbital facility and put off detailed design work until the program was approved. Throughout the approval process, members of the Task Force worked to define the space station in terms of 107 missions rather than a specific design.[9] There were many reasons for delaying design decisions in favor of missions. Some were technical; some were political.

On the technical side, NASA officials could point to one of their internal management studies. A report by NASA center director Donald P. Hearth had fingered premature design work as a major cause of cost growth and missed deadlines.[10] Hearth's report provided a technical rationale for emphasizing missions over design. John Hodge did not want NASA engineers to lock the agency into a detailed design for the space station without first analyzing mission requirements, fearing that a prematurely designed space station would prove unresponsive to the needs of users.

On the political side, NASA officials could recall the process by which the Space Transportation System had been approved some twelve years earlier. During the White House review of NASA's proposal for that system, the actual design of the space shuttle had fallen onto the negotiating table. NASA officials had to negotiate the size and shape of the space shuttle in order to get it approved. They were determined not to let that happen again. By de-emphasizing space station design, they reduced the probability that presidential aides might try to reshape the facility as a condition for White House support.

The de-emphasis of design also helped members of the Task Force build a constituency for the space station, since it focused attention on the unifying issue of missions rather than the divisive issue of design. Potential users got genuinely excited when contemplating missions they might carry out from a space station; they got distressed when they contemplated the contradictions in design. The mission emphasis was especially important in helping the Europeans, Canadians, and Japanese put aside their concerns and become involved during the approval process.

After his retirement from NASA in 1986, John Hodge reminisced about the challenge of building so large a constituency for a space station that would have to serve so many purposes. At one of the many briefings he had given in his effort to tell the space station story, Hodge recalled, he had announced in a totally mischievous way that they had settled on a name for the facility. "I started the whole thing off with the following story and said that we decided after a lot of discussion to call the space station Proteus."

"Now you have to understand Greek history," Hodge explained. "Who was Proteus? Proteus was a god who was told by Zeus or whoever it was that if he was ever caught by a human he had to tell the truth, but in order to avoid telling the truth he could change his shape, so that he couldn't get caught. And so every time a mortal tried to get hold of him he changed his shape into something else, and that's the reason we wanted to call the space station Proteus because you could always change its shape if you wanted to avoid telling the truth."[11]

Define Total Program Costs. Just as NASA officials had to formulate a workable solution to the problem of many users, so they had to fashion an acceptable answer to the questions about program cost. Their reputation as technically competent engineers obliged them to issue a realistic cost estimate. At the same time, they had to keep program costs low enough to get the program approved. Doing both—especially in the absence of a detailed design upon which they could base a firm cost estimate—proved very tricky.

NASA executives had confronted no such problem in 1961 when they presented their cost estimate for the trip to the moon. The political environment was such that NASA officials felt free to issue a realistic cost estimate for the voyage. Their $20- to $40-billion estimate was breathtakingly large by the standards of that day. The federal government as a whole spent $98 billion in fiscal year 1961, so a $20-billion mark—even though it would be spent over eight years—looked like an enormous number to people involved in the decision to go to the moon. The $8-billion space station proposal, though it generated strong resistance, was a paltry sum in a year (1984) when the federal government spent $852 billion. Starting with a realistic cost estimate, NASA managers from the Apollo era found it much easier to meet the cost goal for the lunar voyage. The actual cost of the expedition (through the first flight to the moon) totaled slightly more than $21 billion.[12]

NASA officials faced a very different situation in seeking approval for the Space Transportation System in 1971. They found themselves negotiating the cost of the program as a price for winning White House support. What began as a $10- to $13-billion initiative emerged from the White House as a $5.15-billion program, leaving NASA with a shuttle configuration that many believed was technologically inferior to the two-stage reusable system and a cost estimate that agency managers could not meet.

NASA officials promoting the space station were determined to avoid shuttle-type negotiations over cost and design. If Task Force members had their way, they would avoid a specific cost estimate until the program was approved and under way. Emphasizing their approach to space station design, Task Force leaders observed that it was premature to set a specific cost goal until NASA had completed its phase B definition and preliminary design studies.

This approach was not well received. The people reviewing NASA's proposal might not be design engineers, but they could add up a column of figures. Faced with mounting pressure to reveal the total program cost, NASA officials placed an $8-billion estimate on the facility. The number was, as one Task Force leader observed, simply "a stopping point in an upward march."[13] As James Beggs frequently pointed out, "Space stations are the kind of development that you can buy by the yard," suggesting that NASA could reduce the size of the space station if the estimate proved too low.[14]

To meet the $8-billion cost estimate, NASA officials had to maintain some very restrictive definitions about what the program contained. They had done this to a lesser extent in the past. The agency's much-heralded ability to meet the $20- to $40-billion cost goal for the Apollo program bought only the first landing on the moon, not the five that followed.[15] The $5.15-billion price tag on the Space Transportation System took NASA through the test phase only, buying just two of the five orbiters NASA officials wanted to acquire.[16]

NASA officials omitted from their $8-billion estimate the cost of transporting the space station from earth to orbit, the cost of operating the space station once it was in orbit, and the cost of conducting experiments on it. The $8-billion estimate did not cover the cost of future elements of the space station necessary to carry out all 107 missions planned for the first decade of operations or components for the decades beyond. Other people might include those costs in their definition of the total program, but NASA officials did not. NASA officials successfully maintained that those costs were not appropriately included within the definition of the initial facility.

To fit a multi-purpose space station in an $8-billion estimate, NASA officials had to go even further. Task Force leaders put in a smaller reserve for possible changes than many people wanted, they limited the spending for ground-support equipment, and they substantially restricted the amount of money NASA field centers could spend from program funds to maintain technical capability. Hodge criticized the latter as "totally legitimized theft" and observed in general that "the more money we put in the budget the more they will spend for the same capability."[17] In the long run, Hodge's adversaries within NASA prevailed. New executives took over and members of the Task Force moved on. The new officers added $3.6 billion for ground support, including $1.5 billion to help run the NASA field centers, and $2.5 billion to the cost reserve. They added $600 million for definition costs, a sum originally included within the $8-billion estimate. That, plus a few design changes, pushed the cost estimate for the space station up to $14.5 billion.[18]

Faced with an increasingly restive Congress, James Fletcher (in his second tour of duty as NASA Administrator), scaled back the design of the space station—by the yard. Eliminated from the space station concept as the Task Force had defined it were one of the two unmanned platforms and much of the satellite-servicing capability.[19]

NASA officials understood back in 1983 that a $12- to $14-billion cost estimate would excite an already substantial opposition. Rather than reduce the station itself, they reduced their definition of what the estimate would include. This allowed them to issue a cost estimate that was realistic for the program as they defined it and at the same time small.

When forced to take an incremental approach toward the initiation of new programs, agency officials begin with cost estimates that are low and try to phase the rest in later or pay for them through other accounts.

Debates over costs often turn on disagreements over the scope of the estimate, not the technical accuracy of the estimates themselves. The confusion that results helps obscure an issue that program advocates would prefer remain cloudy anyway.

Results. NASA officials received a very weak commitment to proceed with the space station when Congress and the President approved the program in 1984. The approval was perceived, especially on Capitol Hill, as more of a mandate to start planning than a commitment to finish building. Policy makers reserved the right to appropriate only enough money to advance the program year by year and to reconsider the purpose of the project as circumstances changed.

By agreeing to put off important design decisions for at least two years, NASA officials encouraged an exceptionally long period of political review. In both the shuttle and Apollo programs, design decisions quickly followed presidential approval. NASA officials announced the configuration for the space shuttle in March 1972, less than three months after President Nixon approved that program.[20] NASA employees took just fourteen months following President Kennedy's speech to Congress to devise the method for taking Americans to the moon.[21] No similar commitment followed the space station decision. NASA slogged through a series of designs—a power tower (1984), a dual keel (1985), a baseline configuration derived from the dual-keel design (1986), a revised baseline configuration to be built along a single boom (1987), and a rephased revised baseline configuration that delayed introduction of certain subsystems (1989). Additionally, NASA submitted to Congress the required study on a man-tended space station (1986) and fought off attempts to substitute an industrial space facility (1987–1988).

Politicians, knowing that an appropriation to start construction would imply an end to the period of review, resisted funding decisions that would let NASA "bend metal." Under the funding pattern approved by President Reagan in 1984, NASA officials expected to start full-scale production of the station in 1987 (see Table 1). In fact, full funding for construction was deferred in FY 1987, FY 1988, and again in FY 1989. As a point of contrast, five years after President Kennedy approved the first trip to the moon, NASA engineers and their contractors conducted the first flight test of an unmanned Apollo space capsule from the Kennedy Space Center.[22] Five years after President Reagan approved the space station, NASA officials were waiting for still another presidential go-ahead on space station design.[23]

Had they possessed a long-term objective, NASA officials would have been in a much better position to explain the purpose of the space station and thus to accelerate its design. The lack of such an objective placed NASA professionals in the difficult position of having to defend a pro-

Table 1
Space Station Appropriations, Planned and Actual (Millions of Dollars)

Fiscal Year	(1) NASA letter 9/8/83	(2) What Reagan Approved, 1984	(3) Actual Presidential Request	(4) Actual Congressional Appropriation
1985	225	150	150	155.5
1986	270	250	226	205
1987	1,040	1,250	410	410
1988	2,215	1,700	767	425
1989	2,420	2,000	967	900
1990	1,510			
1991	320			

NOTE: The figures in columns 1 and 2 are stated in 1984 dollars. Those in columns 3 and 4 are stated in current-year dollars. The numbers in column 2 are taken from a graph and are therefore approximations.

gram on grounds other than those for which it had been conceived.

Historically, space enthusiasts had viewed the space station as a critical step in the exploration of space, the essence of NASA's mandate. Since the agency received no such mandate in the two decades following the U.S. landing on the moon, NASA officials had to fall back on short-term goals to justify the station. They had to make do with arguments about space transportation, space manufacturing, and satellite servicing. These were considerably less than the real thing. Lacking an approved long-range objective, NASA officials had to present the space shuttle and space station as ends in themselves. This compromised NASA's ability to offer realistic advice on their purpose, their overall cost, and the technical difficulties involved in building them.

In the absence of a long-range vision for the nation's space program, both Presidents Nixon and Reagan delegated the responsibility for making decisions about specific projects to their White House staff. Since staff members disagreed on the overall purpose of the space program, they could hardly be expected to agree on its details. The shuttle and the space station decisions emerged from the White House policy review process affected by the need for maneuvering, bargaining, and compromise. President Kennedy, by contrast, imposed his preferences on the warring factions. His advisers disagreed on the merits of various alternatives, but they could not escape Kennedy's desire for a space spectacular at which the U.S. could win. As a result, the Apollo decision moved to the Oval Office much quicker and in a much less compromised form.

Formulating science and technology policy in the absence of long-term goals creates a difficult dilemma for the people trying to initiate new programs. Completing complex engineering tasks in projects that take a long time to build requires a sustained political commitment. Yet through the second and third decades of U.S. space exploration, NASA officials had to operate with less of a political commitment than that possessed by agencies with simpler tasks to perform. Even agencies that paid social security benefits or the interest on the national debt operated with a stronger political commitment than NASA did.

In the absence of policy commitments, NASA officials turned toward incremental strategies to get the space station approved. The program that emerged was less than perfect and difficult to implement. Had the people who wanted a space station waited for consensus on space policy to develop, the initiative would have been delayed for many years. "It was a Hobson's choice," said Jim Beggs, reflecting on one aspect of the dilemma. If NASA did not accept the space station under the conditions within which they could get it, they would be left with nothing at all.[24]

In their quest to preserve flexibility and save money, politicians managed to avoid a commitment to a long-range space policy for twenty years. Flexibility has its advantages, allowing policies to shift as new circumstances emerge. In the case of the space station, however, it skewed the decision in ways that would have been unacceptable during the Apollo era. A guiding space policy, imposed on warring factions by a strong president, would have forced policy makers to resolve many of the inconsistencies that weakened the civilian space program after the Apollo era.

NOTES

INTRODUCTION: THE VISION

1. James M. Beggs, "Why the United States Needs a Space Station" (remarks prepared for delivery at the Detroit Economic Club and Detroit Engineering Society, June 23, 1982), NASA History Office; reprinted under the same title in *Vital Speeches* 48 (August 1, 1982): 615–617.

2. George A. Keyworth, "The U.S. Space Program—Where Do We Go from Here?" (proposed remarks to the Nineteenth Joint Propulsion Conference, Seattle, Wash., June 27, 1983), NASA History Office.

3. John Hodge interview, July 10, 1985, NASA History Office.

4. Tom Wolfe, "Columbia's Landing Closes a Circle," *National Geographic* 160 (October 1981): 475–476. See also Tom Wolfe, *The Right Stuff* (New York: Farrar, Straus & Giroux, 1979).

5. Beggs, "Why the United States Needs a Space Station"; Space Task Group, *The Post-Apollo Space Program: Directions for the Future,* Report to the President (Washington, D.C.: Executive Office of the President, September 1969).

6. For the traditional definitions of *station,* see Philip Babcock Gove, ed., *Webster's Third New International Dictionary* (Springfield, Mass.: G. & C. Merriam Co., 1963), p. 2229; Daniel Herman, Workshop on Automated Space Station, Washington, D.C., March 18, 1984, in U.S. Senate, Committee on Appropriations, a Subcommittee, *Department of Housing and Urban Development, and Certain Independent Agencies Appropriations for Fiscal Year 1985,* 98th Cong., 2d sess., 1984, p. 1266.

7. K. E. Tsiolkovskiy, *The Investigation of Universal Space by Means of Reactive Devices* (1911 and 1926), NASA translations in Tsiolkovskiy, *Works on Rocket Technology,* November 1965, NASA History Office; Herman Oberth, *Die Rakete zu den Planetenraumen* (1923), NASA translation, NASA History Office; Guido von Pirquet, "Fahrtrouten," *Die Rakete,* May 1923 to April 1929; Fritz Sykora, "Guido von Pirquet: Austrian Pioneer of Astronautics" (paper presented at the Fourth History Symposium of the International Academy of Astronautics, Constance, German Federal Republic, October 1970), NASA History Office.

8. Oberth, *Rockets in Planetary Space,* pp. 93–97.

9. Quoted from Daniel Lang, "A Reporter At Large: A Romantic Urge," *New Yorker* 27 (April 21, 1951): 74.

10. Wernher von Braun, "Crossing the Last Frontier," *Collier's* 129 (March 22, 1952): 24–29, 72–74.

11. NASA Office of Program Planning and Evaluation, "The Long Range Plan of the National Aeronautics and Space Administration," December 16, 1959, table 1

and p. 28; Minutes of the Research Steering Committee on Manned Space Flight, NASA Headquarters Office, Washington, D.C., May 25–26, 1959, pp. 2, 9; both in NASA History Office.

12. Minutes of the Research Steering Committee on Manned Space Flight, Ames Research Center, June 25–26, 1959, p. 6, NASA History Office; Harry Goett interview, June 17, 1988.

13. Langley Research Center, "A Report on the Research and Technological Problems of Manned Rotating Spacecraft," NASA Technical Note D-1504, August 1962; Institute of Aeronautical Sciences, *Proceedings* of the Symposium on Manned Space Stations, April 20–22, 1960, Los Angeles, Calif. For papers from the 1960 symposium see also *Aero/Space Engineering* 19, no. 5 (May 1960).

14. George M. Low, "Manned Space Flight," *NASA-Industry Program Plans Conference,* July 28–29, 1960 (Washington, D.C.: Government Printing Office, 1960), pp. 79–80.

CHAPTER 1: THE RACE

1. John F. Kennedy, Memorandum for the Vice President, April 20, 1961, NASA History Office. The second quotation is taken from George M. Low, "Manned Space Flight," in *NASA-Industry Program Plans Conference, July 28–29, 1960* (Washington, D.C.: Government Printing Office, 1960), p. 80.

2. Kennedy, Memorandum for the Vice President, April 20, 1961; NASA, "The Long Range Plan," January 12, 1961, p. 4, NASA History Office.

3. The *Washington Post* editorial appeared on p. A18 of its April 13, 1981, issue. The score was given by Congressman David S. King, U.S. House, Committee on Science and Astronautics, *1962 NASA Authorization,* 87th Cong., 1st sess., 1961, pt. 1, p. 375.

4. John F. Kennedy, *The Kennedy Presidential Press Conferences* (New York: Earl M. Coleman Enterprises, 1978), April 21, 1961, p. 87.

5. James Webb, "Administrator's Presentation to the President," March 21, 1961; Agenda for NASA-BOB Conference with the President, March 22, 1961; NASA, Long Range Plan, Manned Space Flight Program, August 8, 1960; all in NASA History Office.

6. NASA, "The Long Range Plan," January 12, 1961, pp. 4, 18.

7. Low, "Manned Space Flight," pp. 80–81. See also George M. Low, Memorandum for Associate Administrator, Subject: Transmittal of Report Prepared by Manned Lunar Working Group, February 7, 1961, p. 3.

8. Webb, "Administrator's Presentation," March 21, 1961.

9. John Logsdon, *The Decision to Go to the Moon* (Cambridge: MIT Press, 1970), p. 99.

10. Quoted from Hugh Sidey, *John F. Kennedy, President* (New York: Atheneum, 1963), p. 122.

11. Office of the Vice President, Memorandum for the President, Subject: Evaluation of Space Program, April 28, 1961, p. 3.

12. Wernher von Braun to the Vice President of the United States, April 29, 1961, p. 1, NASA History Office.

13. NASA, discussion material used with the Vice President, April 22, 1961, p. 1, NASA History Office.

14. H. Hermann Koelle, Erich E. Engler, and John W. Massey, "Design Criteria and Their Application to Economical Manned Satellites," *Aero/Space Engineering* 19 (May 1960): 90; Langley Research Center Staff, *A Report on the Research and Technological Problems of Manned Rotating Spacecraft,* Technical Note D-1504 (Washington, D.C.: NASA, August 1962), p. 13.

15. Von Braun to the Vice President, April 29, 1961.

16. NASA, Memorandum to the Vice President, April 22, 1961.

17. Von Braun to the Vice President, April 29, 1961.

18. NASA, discussion material, April 22, 1961, p. 3.

19. Kennedy, *Presidential Press Conferences,* April 21, 1961, p. 87.

20. Hugh L. Dryden, "Space Technology and the NACA" (address delivered on his behalf at the Twenty-sixth Annual Meeting of the Institute of the Aeronautical Sciences, January 27, 1958), published in *Aeronautical Engineering Review,* March 1958 (quotation on p. 33). George M. Low, Minutes of the Meeting of the Research Steering Committee on Manned Space Flight held at NASA Headquarters Office, Washington, D.C., May 25–26, 1959, p. 9.

21. Logsdon, *Decision to Go to the Moon,* pp. 118–121.

22. James Webb to Jerome Wiesner, May 2, 1961, NASA History Office.

23. Quoted from Logsdon, *Decision to Go to the Moon,* p. 118.

24. David Bell interview conducted by John Logsdon, October 4, 1967, pp. 17, 18; see also David Bell, Memorandum for the President, Subject: NASA Budget Problem, n.d., NASA History Office.

25. Loyd S. Swenson, James M. Grimwood, and Charles C. Alexander, *This New Ocean: A History of Project Mercury,* SP-4201 (Washington, D.C.: NASA, 1966), pp. 275–296.

26. Statement by James E. Webb, May 1, 1961, news release no. 61–96, NASA History Office.

27. NASA, *Pocket Statistics* (Washington, D.C.: NASA, January 1986), p. B14; Tom Wolfe, *The Right Stuff* (New York: Farrar, Straus & Giroux, 1979), p. 78.

28. James E. Webb and Robert S. McNamara, "Recommendations for Our National Space Program: Changes, Policies, Goals," May 8, 1961, p. 13, NASA History Office.

29. James Webb testimony, U.S. House, Committee on Science and Astronautics, *Discussion of Soviet Man-In-Space Shot,* 87th Cong., 1st sess., April 13, 1961, p. 31; Hugh Dryden testimony, U.S. House, Committee on Science and Astronautics, *1962 NASA Authorization,* 87th Cong., 1st sess., July 11–13, 1961, pt. 3, p. 1043; Webb and McNamara, "Recommendations for Our National Space Program," pp. 9–10, 13.

30. The only passing reference to anything resembling a space station is a short statement under the Apollo program budget section which states that Project Apollo could be used as a "multi-manned orbiting laboratory to qualify the spacecraft." Beyond that, the report mentioned absolutely no funding to start work on a space station or laboratory in space. Webb and McNamara, "Recommendations for Our National Space Program," pp. 2–3.

31. Courtney G. Brooks, James M. Grimwood, and Loyd S. Swenson, *Chariots*

for Apollo: A History of Manned Lunar Spacecraft, SP-4205 (Washington, D.C.: NASA, 1979), p. 47 and chap. 3.

32. Army Ballistic Missile Agency, "A Lunar Exploration Based upon Saturn-Boosted Systems," ABMA Report DV-TR-2-60, February 1, 1960, pp. 224–240; Webb and McNamara, "Recommendations for Our National Space Program," p. 3.

33. Wernher von Braun, Ernst Stuhlinger, and Heinz H. Koelle, "ABMA Presentation to the National Aeronautics and Space Administration," ABMA Report D-TN-1-59, December 15, 1958, NASA History Office.

34. Brooks, Grimwood, amd Swenson, *Chariots for Apollo,* p. 59.

35. Wernher von Braun, "Concluding Remarks by Dr. Wernher von Braun about Mode Selection for the Lunar Landing Program," given to Dr. Joseph F. Shea, Deputy Director (Systems), Office of Manned Space Flight, June 7, 1962, p. 4, NASA History Office. The possibility of such an unfortunate accident was raised in the U.S. House, Committee on Science and Astronautics, Subcommittee on Manned Space Flight, *1963 NASA Authorization: Hearings on H.R. 10100 (Superseded by H.R. 11737),* 87th Cong., 2d sess., 1962, pp. 528–529, 810.

36. John Houbolt to Robert C. Seamans, November 15, 1961, NASA History Office.

37. Von Braun, "Concluding Remarks," p. 2.

CHAPTER 2: ONE NEW INITIATIVE

1. NASA, *Current News* (Apollo 11 Special), pts. 1–2, July–August 1969, NASA History Office.

2. Quoted from Theodore C. Sorensen, *Kennedy* (New York: Harper & Row, 1965), p. 180.

3. Richard Nixon, Memorandum for the Vice President et al., February 13, 1969, taken from Space Task Group, *The Post-Apollo Space Program: Directions for the Future,* Report to the President (Washington, D.C.: Executive Office of the President, September 1969).

4. Space Task Group, *The Post-Apollo Space Program.*

5. NASA, *The Post-Apollo Space Program,* summary of NASA's report to the President's Space Task Group, "America's Next Decades in Space" (Washington, D.C.: NASA, September 1969), p. 1.

6. Space Task Group, *The Post-Apollo Space Program,* pp. 13, 15, 20.

7. The Vice President's statement was widely reported when he made it. See *New York Times,* July 17, 1969, pp. 1, 22.

8. Space Task Group, *The Post-Apollo Space Program,* pp. 11, 20.

9. North American Rockwell Space Division, "Space Station Program Phase B Definition," Second Quarterly Progress Report, contract NAS9-9953, MSC, Houston, March 13, 1970, NASA History Office; see also W. Ray Hook, "Historical Review," *Journal of Engineering for Industry: Transactions of the ASME* 106 (November 1984): 276–286.

10. For a general review of these efforts see, John M. Logsdon, "Space Stations: A Policy History," prepared for the Johnson Space Center, NASA, contract NAS9-16461, George Washington University, Washington, D.C., n.d.

11. NASA, "Space Station, Shuttle Task Groups," *NASA News,* release no. 69-70, May 7, 1969.

12. NASA, *The Post-Apollo Space Program,* p. 1.

13. Herbert E. Krugman, "Public Attitudes toward the Apollo Space Program, 1965–1975," *Journal of Communication* 27 (Autumn 1977): 87–93.

14. Wernher von Braun to the Vice President of the United States, April 29, 1961, p. 9, NASA History Office.

15. Robert Mayo to Thomas Paine, July 28, 1969, reported in John Logsdon, "From Apollo to Shuttle: Policy-Making in the Post-Apollo Era" (unpublished manuscript dated Spring 1983), chap. 5, p. 7, NASA History Office.

16. Logsdon, "From Apollo to Shuttle," chap. 5, p. 8.

17. Thomas Paine, NASA Administrator, to President Richard Nixon, December 17, 1969, NASA History Office. For further details, see Logsdon, "From Apollo to Shuttle," chap. 5, p. 29; and Director, History Division, to Associate Administrator for External Relations, Subject: Termination of Saturn Vehicle; Skylab as Space Station, March 23, 1988, NASA History Office. See also NASA, "Saturn Launch Vehicle Curtailment," *NASA News,* release no. 68–139, August 2, 1968; NASA "NASA Interim Operation Plan," *NASA News,* release no. 68-141, August 8, 1968; and Dr. Thomas O. Paine, "NASA Future Plans News Conference," *NASA News,* January 13, 1970, p. 3; all in NASA History Office.

18. Memorandum from Peter Flanagan to Thomas O. Paine and Robert Mayo, January 6, 1970, NASA History Office.

19. Richard M. Nixon, "The Future of the United States Space Program," *Weekly Compilation of Presidential Documents,* March 7, 1970, p. 329.

20. For a history of the complex legislative struggle, see Logsdon, "From Apollo to Shuttle," chap. 5, pp. 43–53.

21. An elaborate history of the orbital workshop program can be found in W. David Compton and Charles D. Benson, *Living and Working in Space: A History of Skylab,* SP-4208 (Washington, D.C.: NASA, 1983); see especially pp. 52–56 and 114–118.

22. James Fletcher interview, October 14, 1985.

23. President's Science Advisory Committee, *The Next Decade in Space* (Washington, D.C.: Executive Office of the President, Office of Science and Technology, March 1970), pp. 35–36, 38, 47, 49–50.

24. Fletcher interview, October 14, 1985; Space Task Group, *The Post-Apollo Space Program,* p. 20.

25. Fletcher interview, October 14, 1985.

26. E. P. Smith, "Space Shuttle in Perspective: History in the Making" (paper delivered at the American Institute of Aeronautics and Astronautics' Eleventh Annual Meeting and Technical Display, Washington, D.C., February 24–26, 1975), NASA History Office.

27. John Logsdon, "The Decision to Develop the Space Shuttle," *Space Policy* 2 (May 1986): 110.

28. The two-stage, fully reusable system would cost more than $10 billion and require NASA to fly 445 missions over a ten-year period in order to make the system cost-effective. Smith, "Space Shuttle," p. 10. The conclusions of the Mathematica study can be found in Klaus P. Heiss and Oskar Morgenstern, *Economic Analysis of*

the Space Shuttle System: Executive Summary, National Aeronautics and Space Administration contract NASW-2081, January 31, 1972, and Klaus P. Heiss and Oskar Morgenstern, "Factors for a Decision on a New Reusable Space Transportation System," Memorandum for Dr. James C. Fletcher, Administrator, NASA, Mathematica, October 28, 1971. Both of these summaries can be found in the NASA History Office files. The Mathematica experts substituted the "mission model" for NASA's earlier and less sophisticated efforts at determining cost effectiveness by cost per pound. Given the fixed costs of ground support and vehicle hardware, cost per pound on a space launch varied enormously depending upon the frequency with which people on the ground could organize a launch and their ability to reuse parts of the launch vehicle.

29. Fletcher interview, October 14, 1985; K. Heiss interview conducted by John M. Logsdon, November 13, 1975, and quoted in Logsdon, "The Space Shuttle Program: A Policy Failure?" *Science* 232 (May 20, 1986): 1101. See also Claude E. Barfield, "Technology Report: Intense Debate, Cost Cutting Precede White House Decision to Back Shuttle," *National Journal* 4 (August 12, 1972): 1289–1299; John Logsdon, "The Space Shuttle Decision: Technological and Political Choice," *Journal of Contemporary Business* 7, no. 3 (1978): 13–30; and John Logsdon, "The Decision to Develop the Space Shuttle."

30. Logsdon, "The Space Shuttle Program," p. 1103.

31. Ibid., pp. 1102–1104; Claude E. Barfield, "Technology Report: NASA Broadens Defense of Space Shuttle to Counter Critics' Attacks," *National Journal* 4 (August 19, 1972): 1325–1327.

32. Fletcher interview, October 14, 1985. See also NASA photos 72-H-30, 32, and 38 in the NASA History Office.

33. George Low, Memorandum for the Record, Subject: Meeting with the President on January 5, 1972, dated January 12, 1972, p. 2, NASA History Office.

34. Office of the White House Press Secretary (San Clemente, California), The White House, Statement by the President, January 5, 1972, NASA History Office; "Space Shuttle Program," Statement by the President Announcing the Decision to Proceed with Development of the New Space Transportation System, *Weekly Compilation of Presidential Documents,* January 5, 1972, pp. 27–28.

35. Office of the White House Press Secretary (San Clemente, California), The White House, Press Conference of Dr. James Fletcher and George M. Low, San Clemente Inn, California, January 5, 1972, p. 1, NASA History Office.

36. The White House, Statement by the President, January 5, 1972, p. 1.

37. The White House, Press Conference of Fletcher and Low, January 5, 1972, p. 8.

38. W. R. Lucas, Program Development Memorandum to Dr. Rees, June 16, 1970, NASA History Office; "NASA to Seek $415–465 Million for Shuttle in FY '73," *Space Daily,* July 1, 1971, p. 3; Claude E. Barfield, "Space Report: NASA Gambles Its Funds, Future on Reusable Space Shuttle Program," *National Journal* 3 (March 13, 1971): 551; Logsdon, "The Space Shuttle Program," p. 1102; NASA, "Space Shuttle Decisions," *NASA News,* release no. 72-61, March 15, 1972; NASA, "Response to NBC Inquiry on Shuttle and Other Manned Space Flight Program Costs," March 10, 1981, NASA History Office.

39. Fletcher interview, October 14, 1985.

40. Ibid.; J. K. Davies, "A Brief History of the Voyager Project: The End of the Beginning," *Spaceflight* 23 (February 1981): 35–41.

CHAPTER 3: BEGGS

1. James Fletcher interview, October 14, 1985; James M. Beggs interview, September 3, 1985; James C. Fletcher to the Honorable Caspar W. Weinberger, Deputy Director, Office of Management and Budget, March 6, 1972, NASA History Office.

2. Beggs interview, September 3, 1985.

3. U.S. Executive Office of the President, *America's New Beginning: A Program for Economic Recovery* (Washington, D.C.: Executive Office of the President, February 18, 1981), chap. 6, pp. 35–36.

4. Beggs interview, September 3, 1985.

5. James Montgomery Beggs, Résumé, n.d. (but prepared about 1980).

6. Beggs interview, September 3, 1985.

7. "NASA Decides on Twenty-six Changes in Shuttle Program to Save $360 Million," *Defense/Space Daily,* December 27, 1974, p. 288; "Space Shuttle Design Changes Cut Cost," *Aviation Week & Space Technology,* November 13, 1972, p. 18. See also the file in the NASA History Office labeled "Shuttle Costs."

8. For the target dates, see NASA, *Astronautics and Aeronautics, 1972* (Washington, D.C.: Government Printing Office, 1974), pp. 383–384; and MSF Schedule Assessment of Major Space Shuttle Milestones, January 31, 1975, NASA History Office. Information on the orbiter's main engines can be found in NASA, "The Space Shuttle Main Engine and the Solid Rocket Booster" (briefing conducted by James R. Thompson, George B. Hardy, and John Taylor, Marshall Space Flight Center, Huntsville, Alabama, October 14, 1980), NASA History Office. See also "Engine Explosion May Delay Launch of Space Shuttle," *Washington Post,* January 4, 1979, p. A-2.

9. NASA Daily Activities Report, "Space Transportation Systems: Thermal Protection System," Tuesday, March 13, 1979, NASA History Office.

10. M. E. Merrell, National Space Transportation System, Critical Items List, March 1986, NASA History Office.

11. Beggs interview, September 3, 1985.

12. NASA, General Management Status Report (GMSR), May 18, 1981, STS-1 file, NASA History Office.

13. Craig Covault, "NASA Studies Cuts to Shuttle Costs," *Aviation Week & Space Technology,* August 17, 1981, pp. 27–29.

14. Beggs interview, September 3, 1985.

15. Report of the NASA Transition Team, George M. Low, Team Leader, to Mr. Richard Fairbanks, Director, Transition Resources and Development Group, December 19, 1980, NASA History Office.

16. Beggs interview, September 3, 1985.

17. U.S. Senate, Committee on Commerce, Science, and Transportation, *Nominations—NASA,* 97th Cong., 1st sess., June 17, 1981, p. 1.

18. Hans Mark, *The Space Station: A Personal Journey* (Durham, N.C.: Duke University Press, 1987).

19. Senate Committee on Commerce, Science, and Transportation, *Nominations*, p. 22.

20. Beggs interview, September 3, 1985.

21. Senate Committee on Commerce, Science, and Transportation, *Nominations*, p. 22.

CHAPTER 4: THE TEAM

1. John Hodge interview, July 10, 1985.

2. John Hodge interview, June 27, 1986.

3. Ibid. See also James Dow, *The Arrow* (Toronto: James Lorimer & Co., 1979), pp. 129–131; and Murray Peden, *Fall of an Arrow* (Stittsville, Canada: Canada's Wings, 1978).

4. Hodge interview, June 27, 1986.

5. John Hodge interview, September 16, 1985.

6. Hodge interviews, July 10, 1985, and June 27, 1986.

7. Philip Culbertson interview, September 14, 1987.

8. Hodge interview, July 10, 1985.

9. Philip Culbertson interview, July 15, 1985.

10. Hodge interview, June 27, 1986.

11. Ibid.

12. Terence Finn interview, December 31, 1985.

13. Ibid. See also "Boland Subcommittee Approves All Added Funds for Shuttle," *Defense/Space Business Daily* 104 (May 17, 1979): 85–86; and NASA, Associate Administrator/Comptroller, Budget Operations Division, Chronological History of the FY 1980 Budget Submission, NASA History Office.

14. Terence T. Finn interview conducted by Sylvia D. Fries, June 12, 1985, biography file, NASA History Office.

15. Hodge interview, June 27, 1986.

16. Daniel H. Herman interview conducted by Sylvia D. Fries, March 26, 1985, biography file, NASA History Office.

17. Ibid.

18. Daniel Herman interview, July 19, 1985.

19. Herman interview, March 26, 1985.

20. Herman interview, July 19, 1985.

21. Hodge interview, June 27, 1986.

22. Robert Freitag interview conducted by Sylvia Fries, May 16, 1985, biography file, NASA History Office.

23. Ibid.

24. "Robert F. Freitag," in Shirley Thomas, *Men of Space*, vol. 7 (Philadelphia: Chilton Co., 1965), p. 37. See also Harvey M. Sapolsky, *The Polaris System Development* (Cambridge, Mass.: Harvard University Press, 1972).

25. Thomas, *Men of Space*, pp. 38–40.

26. Freitag interview, May 16, 1985. See also Capt. Charles W. Styer and Com.

Robert F. Freitag, "The Navy in the Space Age," *U.S. Naval Institute Proceedings* 86 (March 1960): 86–93: and Robert F. Freitag, "The Effect of Space Operations on Naval Warfare," in *Naval Review, 1962–1963,* ed. Frank Uhlig (Annapolis, Md.: U.S. Naval Institute, 1962).

27. Freitag interview, May 16, 1985.

28. Robert Freitag interview, December 30, 1985.

29. Freitag interview, May 16, 1985.

30. Robert Freitag interview, July 10, 1985; Robert Freitag, "NASA Philosophy Concerning Space Stations as Operations Centers for Construction and Maintenance of Large Orbiting Energy Systems," *Journal of the British Interplanetary Society* 30 (July 1977): 265; Space Station Task Force roundtable interview, August 4, 1987.

31. Hodge interview, July 10, 1985.

32. [Robert] Freitag to Abe [James Abrahamson] et al., "Space Station Task Force," March 30, 1982. See also Robert F. Freitag to Howard McCurdy, February 7, 1987; both in NASA History Office.

33. Culbertson interview, July 15, 1985.

34. James M. Beggs, Special Announcement, "Establishment of a Space Station Task Force," May 20, 1982, NASA History Office.

35. Hodge interview, June 27, 1986.

36. Craig Covault, "NASA Head Backs Space Station Focus," *Aviation Week & Space Technology,* July 27, 1981, pp. 23–25; James Beggs interview, October 30, 1987.

CHAPTER 5: INDEPENDENCE DAY

1. Ralph Jackson and Roger Barnicki interview, April 18, 1986. See also Craig Covault, "Shuttle Lands on Hard Surface Runway," *Aviation Week & Space Technology,* July 12, 1982, pp. 22–23.

2. U.S. House, Committee on Appropriations, Subcommittee on HUD-Independent Agencies, *Department of Housing and Urban Development–Independent Agencies Appropriations for 1984,* 98th Cong., 1st sess., 1983, pt. 3, p. 20, and pt. 6, p. 68.

3. Hans Mark interview, November 11, 1987. See also Hans Mark, *The Space Station: A Personal Journey* (Durham, N.C.: Duke University Press, 1987), pp. 133–136.

4. Hugh L. Dryden, "Space Technology and the NACA" (speech prepared for delivery at the Institute of Aeronautical Sciences and printed in *Aeronautical Engineering Review,* March 1958); U.S. Senate, Committee on Aeronautical and Space Sciences, NASA Authorization Subcommittee, *NASA Supplemental Authorization for Fiscal Year 1959,* 86th Cong., 1st sess., February 19–20, 1959, p. 46.

5. James Beggs interview, September 3, 1985. See also NASA, Mission Report, "STS-4 Test Mission Simulates Operational Flight—President Terms Success 'Golden Spike' in Space," MR-004, n.d.

6. Beggs interview, September 3, 1985.

7. Rockwell International, "Press Information: Space Shuttle Transportation System," January 1984; U.S. House, Committee on Science and Technology, *Flight*

of STS-4 with Astronauts Captain T. Ken Mattingly and Henry Hartsfield, 97th Cong., 2d sess., July 28, 1982; Craig Covault, "Shuttle Reentry Tests Vehicle Crossrange," *Aviation Week & Space Technology,* July 12, 1982, pp. 20–21; William B. Scott, "Shuttle Lands on Hard Surface Runway," *Aviation Week & Space Technology,* July 12, 1982, pp. 22–23; Henry Hartsfield interview, November 9, 1987.

8. Jackson and Barnicki interview, April 18, 1986. Many details of the President's visit can be confirmed from photographs at the NASA History Office and the Dryden Center Public Affairs Office.

9. Ronald Reagan, "National Initiative on Technology and the Disabled, Remarks at a White House Ceremony," *Weekly Compilation of Presidential Documents,* December 3, 1985, p. 1448; Ronald Reagan, "National Aeronautics and Space Administration, Remarks at the Twenty-fifth Anniversary Celebration," *Weekly Compilation of Presidential Documents,* October 19, 1983, p. 1462.

10. House Committee on Science and Technology, *Flight of STS-4,* p. 15.

11. CBS News Television Broadcasts, "Special Report: The Landing of Space Shuttle IV," Sunday, July 4, 1982.

12. House Committee on Science and Technology, *Flight of STS-4,* p. 15.

13. Rockwell International, "Press Information," 1984, p. 480.

14. CBS News Television Broadcasts, "Special Report," July 4, 1982, pp. 6–7.

15. Jackson and Barnicki interview, April 18, 1986.

16. House Committee on Science and Technology, *Flight of STS-4,* p. 16.

17. Ibid., p. 15; Hartsfield interview, November 9, 1987.

18. Office of the White House Press Secretary, The White House, Fact Sheet, U.S. Civil Space Policy, October 11, 1978, p. 1, NASA History Office.

19. Administration of Ronald Reagan, "United States Space Policy: Fact Sheet Outlining the Policy," *Weekly Compilation of Presidential Documents,* July 4, 1982, pp. 875–876.

20. Ibid., pp. 872–874.

21. Administration of Ronald Reagan, "United States Space Policy: Remarks on the Completion of the Fourth Mission of the Space Shuttle Columbia," *Weekly Compilation of Presidential Documents,* July 4, 1982, p. 871.

22. Memoranda to Hans Mark, James Beggs, and Philip Culbertson from NASA Center Directors, April–May 1982, NASA History Office. See also Mark, *Space Station,* p. 149 and app. 7.

23. Beggs interview, September 3, 1985; NBC Television Network News Archives, "Today" with Jane Pauley, Bryant Gumbel, et al., March 22, 1982, p. 20.

24. James M. Beggs to Edwin Meese, May 21, 1982; James M. Beggs, "Why the United States Needs a Space Station" (remarks prepared for delivery at the Detroit Economic Club and Detroit Engineering Society, June 23, 1982); both in NASA History Office.

25. "Twenty-five Years of the Presidency," statement by Richard Cheney, PBS-TV, June 3, 1986, recorded at the University of California at San Diego.

26. "Twenty-five Years of the Presidency," statement by Jack Watson, PBS-TV, June 3, 1986.

27. "Twenty-five Years of the Presidency," comment by Theodore Sorenson, PBS-TV, June 3, 1986.

28. Mark, *Space Station,* pp. 148–150; James Beggs interview, October 30, 1987;

Hans Mark interview, November 11, 1987; Gilbert D. Rye interview, September 13, 1985.

29. Jackson and Barnicki interview, April 18, 1986.

30. Administration of Ronald Reagan, "United States Space Policy," p. 869; Ronald Reagan, "Text of [Prepared] Remarks of the President on the Landing of the Space Shuttle Columbia," Dryden Flight Research Facility, July 4, 1982, NASA History Office.

31. Administration of Ronald Reagan, "United States Space Policy," p. 870 (emphasis added).

32. Ibid.

33. Mark, *Space Station,* p. 149 and app. 7.

34. Ibid., pp. 149–151; Rye interview, September 13, 1985; James Beggs interview, December 29, 1989.

35. Philip E. Culbertson interview, July 15, 1985.

CHAPTER 6: BUDGET STRATEGY

1. U.S. House, Committee on Appropriations, Subcommittee on HUD-Independent Agencies, *Department of Housing and Urban Development–Independent Agencies Appropriations for 1985,* 98th Cong., 2d sess., March 27, 1984, pt. 6, p. 2.

2. NASA, *Aeronautics and Space Report of the President, 1985 Activities* (Washington, D.C.: NASA, 1987), p. 134.

3. Hugh Heclo, "Issue Networks and the Executive Establishment," in *The New Political System,* ed. Anthony King (Washington, D.C.: American Enterprise Institute for Public Policy Research, 1978), pp. 87–124.

4. Aaron Wildavsky, *The New Politics of the Budgetary Process* (Glenview, Ill.: Scott, Foresman & Co., 1988).

5. Jack Young interview, January 16, 1986.

6. James M. Beggs interview, September 3, 1985.

CHAPTER 7: WHEELS, CANS, AND MODULES

1. At least two programs could be said to have "accomplished" the "laboratory in space" aspect of the space station mission—the long-duration Gemini and Apollo earth-orbital flights, which carried out a variety of space-based experiments, and the Skylab program.

2. Jerome Agee, *The Making of Kubrick's 2001* (New York: New American Library, 1970).

3. Wernher von Braun, "Crossing the Last Frontier," *Collier's* 129 (March 22, 1952): 29; W. Ray Hook, "Historical Review," *Journal of Engineering for Industry: Transactions of the ASME* 106 (November 1984): 277–278.

4. Advertisement, *U.S. Naval Institute Proceedings,* February 1961, p. 9.

5. Maxime A. Faget and Edward H. Olling, "A Summary of NASA Manned Spacecraft Center Advanced Earth Orbital Missions Space Station Activity from 1962

to 1969," NASA, "Compilation of Papers Presented at the Space Station Technology Symposium," Langley Research Center, February 11–13, 1969, pp. 43–98; Lockheed-California, "Study of a Rotating Manned Orbital Space Station," Final Report, vol. 11, Summary, contract NAS9-1665, MSC, Houston, March 1964.

6. Arthur C. Clarke, *2001: A Space Odyssey* (New York: New American Library, 1968), pp. 49, 53.

7. See Gerard K. O'Neill, *The High Frontier: Human Colonies in Space* (New York: William Morrow & Co., 1977); T. A. Heppenheimer, *Colonies in Space* (Harrisburg, Pa.: Stackpole Books, 1977); Richard D. Johnson and Charles Holbrow, *Space Settlements: A Design Study* (Washington, D.C.: NASA, 1977); Steward Brand, ed., *Space Colonies* (Sausalito, Calif.: A CoEvolution Book, published by the Whole Earth Catalog, 1977); and NASA photo, "Color View of Space Colony Interior," photos 75-HC-470 and 75-H-823, August 26, 1975, NASA History Office.

8. See W. David Compton and Charles D. Benson, *Living and Working in Space: A History of Skylab,* SP-4208 (Washington, D.C.: NASA, 1983), pp. 13–14; and Douglas Aircraft Company, "Evaluation of the Usefulness of the MOL to Accomplish Early NASA Mission Objectives," vol. 1, "Summary," contract NAS9-6798, MSC, Houston, October 1967, pp. 4–8, NASA History Office.

9. W. Ray Hook, "Historical Review," *Journal of Engineering for Industry: Transactions of the ASME* 106 (November 1984): 279; and Ray Hook interview, August 23, 1985. On the evolution of space station design, see Adam Louis Gruen, "The Port Unknown" (a dissertation submitted in partial fulfillment of the requirements for the degree of Doctor of Philosophy in the Department of History in the Graduate School of Duke University, 1989); John M. Logsdon, "Space Stations: A Policy History" (prepared for the Johnson Space Center, NASA, contract NAS9-16461, George Washington University, Washington, D.C., n.d.); Sylvia D. Fries, "Space Station: Evolution of a Concept," February 1984, NASA History Office; and Sylvia D. Fries, "2001 to 1992: Political Environment and the Design of NASA's Space Station System," *Technology and Culture* 29 (July 1988): 568–593.

10. Douglas Aircraft Company, "Evaluation of the Usefulness of the MOL.

11. Douglas Aircraft Company, "Report on a System Comparison and Selection Study of a Manned Orbital Research Laboratory," vol. 1, "Technical Summary," contract NAS1-2974, LRC, Hampton, Va., September 1963, pp. 84–85, NASA History Office.

12. Curtis Peebles, "The Manned Orbiting Laboratory," a three-part article in *Spaceflight* 22 (April 1980): 155–160, 22 (June 1980): 248–253, and 24 (June 1982): 274–277. See also Paul B. Stares, *The Militarization of Space: U.S. Policy, 1945–1984* (Ithaca, N.Y.: Cornell University Press, 1985); and Curtis Peebles, *Battle for Space* (New York: Beaufort Books, 1983).

13. NASA Office of Manned Space Flight, Advanced Manned Missions Program, "Space Station Summary Report," June 1969, NASA History Office; Hook, "Historical Review," p. 281; Space Task Group, *The Post-Apollo Space Program: Directions for the Future,* Report to the President, (Washington, D.C.: Executive Office of the President, September 1969), p. 14.

14. Compton and Benson, *Living and Working in Space.*

15. See Nicholas L. Johnson, *The Soviet Year in Space, 1986* (Colorado Springs: Teledyne Brown Engineering, 1987), pp. 53–65.

16. Logsdon, "Space Stations," chap. 2, pp. 29–30. See also NASA, Manned Spacecraft Center, "Statement of Work for Phase B Extension: Modular Space Station Program Definition," November 16, 1970; and Charles Donlan, Space Station Briefing to Robert Seamans, Administrator, NASA, December 19, 1966; both in NASA History Office.

17. Marshall Space Flight Center and McDonnell Douglas Astronautics Company, "Space Station," Space Station Program Extension Period Final Performance Review, contract NAS8-25140, MSFC, Huntsville, Alabama, November 1971; North American Rockwell, Space Division, "Modular Space Station Phase B Extension," First Quarterly Review, contract NAS9-9953, MSC, Houston, May 6, 1971.

18. *The Playthings Directory: The Complete Directory of the American Toy Industry* (New York: McCready Publishing Co., 1954), p. 267. Tinkertoys were first introduced at the American Toy Fair in 1913. For the 503-foot-wide "dual keel," see NASA, "NASA Announces Baseline Configuration for Space Station," *NASA News*, release no. 86-61, May 14, 1986.

19. NASA officials decided to "go modular" in 1966; see Donlan, Space Station Briefing to Robert Seamans, December 19, 1966. For the persistence of the artificial-gravity requirement, see, for example, NASA, [Office of Manned Space Flight], "Statement of Work, Space Station Program Definition (Phase B)," April 14, 1969, chap. 1, p. 8, NASA History Office.

20. M. Mitchell Waldrop, "Space City: 2001 It's Not," *Science 83* 4 (October 1983): 60, 62, and cover.

21. For a list of the thirty-seven Johnson Center studies, see Faget and Olling, "A Summary of NASA Manned Spacecraft Center Advanced Earth Orbital Missions Space Station Activity from 1962 to 1969."

CHAPTER 8: CONFIGURATIONS

1. McDonnell Douglas Astronautics Company, "Evolutionary Space Platform Concept Study," vol. 1, Executive Summary, contract NAS8-33592, MSFC, Huntsville, Alabama, May 1982, NASA History Office.

2. William Marshall interview, November 18, 1985; W. David Compton and Charles D. Benson, *Living and Working in Space: A History of Skylab*, SP-4208 (Washington, D.C.: NASA, 1983), pp. 75, 377–378.

3. Marshall interview, November 18, 1985; see also John M. Logsdon, "Space Stations: A Policy History," prepared for the Johnson Space Center, NASA, contract NAS9-16461, George Washington University, n.d., chap. 3, pp. 7–9.

4. Marshall interview, November 18, 1985; see also Lockheed Missiles and Space Company, "Twenty-five KW Power Module Evolution Study," Final Review Presentation, contract NAS8-32928, MSFC, Huntsville, Alabama, November 28, 1978.

5. Marshall interview, November 18, 1985; McDonnell Douglas, "Evolutionary Space Platform Concept Study," vol. 1.

6. Marshall interview, November 18, 1985.

7. See Maxime A. Faget and Edward H. Olling, "A Summary of NASA Manned Spacecraft Center Advanced Earth Orbital Missions Space Station Activity from 1962

to 1969," in NASA, "Compilation of Papers Presented at the Space Station Technology Symposium, Langley Research Center, February 11–13, 1969," pp. 43–98; see also Logsdon, "Space Stations," chap. 2, p. 14.

8. Allen J. Louviere interview, August 13, 1985.

9. Clarke Covington interview, August 12, 1985.

10. Louviere interview, August 13, 1985.

11. Clarke Covington and Robert O. Piland, "Space Operations Center: The Next Goal for Manned Space Flight?" *Astronautics & Aeronautics* 18 (September 1980): 32; see also The Boeing Company, "Space Operations Center," Systems Analysis, Final Briefing, contract NAS9-16151, JSC, Houston, June 25, 1981.

12. Covington interview, August 12, 1985.

13. Ibid.

14. Robert O. Piland interview, July 24, 1985.

CHAPTER 9: THE FIRST MOVE

1. Memorandum from Carolyn Townsend to Mr. Culbertson, Mr. Hodge, and Capt. Freitag, Office of Space Station, Index of Presentation and Document Files, December 10, 1985. This and the briefing papers are available in the NASA History Office.

2. Terence Finn interview, June 27, 1985; Meeting Record, Space Station Briefing to Industry, March 31, 1982, NASA History Office.

3. NASA, "Proceedings of Space Station Planning Workshop Held at the NASA/Michoud Assembly Facility in New Orleans, Louisiana, November 18–20, 1981," "Introduction," p. 1, NASA History Office; see also Mark B. Nolan interview, August 14, 1985.

4. William Marshall interview, November 18, 1985.

5. NASA, "Proceedings of Space Station Planning Workshop, November 18–20, 1981," "Conclusions and Recommendations," p. 2.

6. Ibid., "Mission Models," p. 1.

7. James Fletcher interview, October 14, 1985.

8. James Fletcher, Final Report, Study Group on Space Station, transmitted to the Administrator of NASA, letter of transmittal from Philip Culbertson to James Fletcher dated November 4, 1982, p. 3, NASA History Office.

9. Fletcher interview, October 14, 1985. See also Marshall interview, November 18, 1985.

10. The first comment is from the Fletcher interview, October 14, 1985. The last two quotations are taken from Fletcher, Final Report, Study Group on Space Station, p. 2.

11. Fletcher interview, October 14, 1985.

12. The suggestion was made by Professor James Arnold. For NASA's response, see John F. Murphy to the Honorable George C. Wortley, House of Representatives, November 1, 1983; and NASA, "Shuttle Expendable Tank May Become Space Platform," *NASA News*, release 77-42, March 7, 1977. Quotations are taken from Fletcher, Final Report, Study Group on Space Station, p. 5.

13. Viewgraph labeled "Some Space Station Budgetary Assumptions," in John

D. Hodge, Space Station Presentation to the NASA Advisory Council, July 21, 1982.

14. M. Mitchell Waldrop, "NASA Wants a Space Station," *Science* 217 (September 10, 1982): 1019.

15. Brian Pritchard interview, August 23, 1985.

16. Donald P. Hearth, NASA Project Management Study, Final Oral Report and Notes on Conclusions and Recommendations to Accompany the Briefing Charts on the NASA Project Management Study, January 21, 1981, pp. 1–2, NASA History Office.

17. Quoted from M. Mitchell Waldrop, "NASA Wants a Space Station," *Science* 217 (September 10, 1982): 1019, and repeated in Waldrop, "Space City: 2001 It's Not," *Science 83* 4 (October 1983): 67. For confirmation, see John Hodge interview, June 27, 1986.

18. Philip Culbertson interview, September 14, 1987.

19. Hodge interview, June 27, 1986.

20. John Hodge interview, July 10, 1985.

21. NASA, "NASA Selects Contractors for Space Station Studies," *NASA News*, release 82-121, August 9, 1982, p. 2, NASA History Office.

22. E. E. Schattschneider, *The Semi-sovereign People* (New York: Holt, Rinehart & Winston, 1960), p. 68.

23. NASA Office of Space Station, *The Space Station: A Description of the Configuration Established at the Systems Requirements Review (SRR)* (Washington, D.C.: NASA, June 1986), pp. 2–3.

24. "Space Station Planning: Acquisition," in John Hodge, Presentation on Space Station Planning to the NASA Center Directors, September 22, 1982, NASA History Office; see also U.S. House, Committee on Science and Technology, Subcommittee on Space Science and Applications, *NASA's Space Station Activities*, 98th Cong., 1st sess., August 2, 1983, p. 23.

25. Fletcher interview, October 14, 1985.

26. Space Station Task Force roundtable interview, August 4, 1987. For the first appearance of the boundary conditions, see John Hodge, Presentation on Space Station Planning to the NASA Center Directors, September 22, 1982, NASA History Office. See also Hodge interview, July 10, 1985.

27. See Terence Finn interview conducted by Sylvia Fries, June 12, 1985; and John Hodge, Space Station Presentation to the NASA Advisory Council, July 21, 1982, viewgraph labeled "Space Station—Reasons Why."

28. Hodge, Presentation on Space Station Planning to the NASA Center Directors, September 22, 1982.

29. NASA, *The Post-Apollo Space Program: Directions for the Future*, Summary of NASA's report to the President's Space Task Group, "America's Next Decades in Space" (Washington, D.C.: NASA, September 1969), p. 1.

30. Administration of Ronald Reagan, "United States Space Policy: Fact Sheet Outlining the Policy," *Weekly Compilation of Presidential Documents*, July 4, 1982, p. 873.

31. See Hodge interview, July 10, 1985; and John Hodge and Richard Carlisle interview conducted by Sylvia Fries, March 4, 1985.

32. See Terence Finn interview conducted by Sylvia Fries, June 12, 1985; Daniel H. Herman interview conducted by Sylvia Fries, March 26, 1985; Daniel Herman

interview, July 19, 1985; and Hans Mark, *The Space Station: A Personal Journey* (Durham, N.C.: Duke University Press, 1987), pp. 139–140.

33. Workshop on Automated Space Station, Washington, D.C., March 18, 1984, from U.S. Senate, Committee on Appropriations, a Subcommittee, *Department of Housing and Urban Development, and Certain Independent Agencies Appropriations for Fiscal Year 1985*, 98th Cong., 2d sess., 1984, p. 1266.

CHAPTER 10: HOW TO ORGANIZE A TASK FORCE

1. Robert Freitag interview conducted by Sylvia Fries, July 12, 1985, NASA History Office.
2. Frank Hoban interview, March 31, 1986.
3. Freitag interview conducted by Sylvia Fries, July 12, 1985.
4. Erasmus H. Kloman, *NASA: The Vision and the Reality* (Washington, D.C.: National Academy of Public Administration, 1985), p. 9.
5. See McDonnell Douglas Astronautics Company, "Evolutionary Space Platform Concept Study," vol. 1, "Executive Summary," contract NAS8-33592, MSFC, Huntsville, Alabama, May 1982; and The Boeing Company, "Space Operations Center," Systems Analysis, Final Report, vol. 1, contract NAS9-16151, JSC, Houston, January 1982; both in NASA History Office.
6. Freitag interview, July 12, 1985. See also William Marshall interview, November 18, 1985; and Clarke Covington interview, August 12, 1985.
7. Donna Pivirotto interview, April 16, 1986.
8. Freitag interview, July 12, 1985.
9. John Hodge interview, July 10, 1985.
10. See NASA, Space Station Task Force, "Space Station Program Description Document," bk. 2, "Mission Description Document," March 1984, introduction, NASA History Office.
11. See ibid., chap. 3, p. 101, and chap. 6, p. 10; and NASA, Lyndon B. Johnson Space Center, "Space Station Program Description Document," bk. 3, "System Requirements and Characteristics," 1st ed., November 1982, chap. 3, pp. 5–6; both in NASA History Office.
12. NASA, Space Station Task Force, "Space Station Systems Definition," bk. 5, lst ed., November 1982, NASA History Office; see also Cecil Gregg interview, November 19, 1985.
13. NASA, John F. Kennedy Space Center, Space Station Operations Working Group, "Space Station Operations Study Plans," May 1, 1983, NASA History Office; see also Frank Bryan and Tom Walton interview, November 4, 1985.
14. Jerry Craig interview, August 12, 1985.
15. Freitag interview, July 12, 1985.
16. Claiborne R. Hicks interview, July 17, 1985.
17. Freitag interview, July 12, 1985. See also James M. Romero interview, August 16, 1985; Mark B. Nolan interview, August 14, 1985; and Francis T. Hoban, "The Space Station Task Force: A Study of a Management Style," October 14, 1987, and Memorandum to Sheahan et al., "What Made the Task Force Work," January 21, 1986, both in NASA History Office.

18. See John Hodge, Presentation on Space Station Planning to NASA Center Directors, September 22, 1982; and Robert Freitag, Presentation to Fletcher Committee, September 2, 1982; both in NASA History Office. See also Clay Hicks interview, July 17, 1985.

19. See John Hodge, "Space Station Presentation to OMB," September 27, 1982, NASA History Office.

20. NASA, Space Shuttle Task Group, L. E. Day, Manager, "NASA Space Shuttle Summary Report," rev. ed., July 31, 1969.

21. Hodge interview, July 10, 1985.

22. John Hodge interview, June 27, 1986. See also Space Station Task Force, "Core Space Station Viewgraphs," July 1982, NASA History Office.

23. See, for example, Presentation made by James Beggs to Ronald Reagan and the Cabinet Council on Commerce and Trade, December 1, 1983, NASA History Office.

24. See, for example, Terence T. Finn, Space Station Task Force, NASA Headquarters, "Presentation on Space Station Planning, Public Affairs Meeting, N.S.T.L.," September 29, 1982, NASA History Office.

25. Terence T. Finn, Note to Howard McCurdy, November 5, 1987, NASA History Office.

26. John D. Hodge, "Space Station Presentation to the NASA Advisory Council," July 21, 1982; Summary Minutes of the NASA Advisory Council, July 21, 1982, NASA Headquarters, Washington, D.C., dated September 13, 1982; both in NASA History Office.

27. Freitag interview, July 12, 1985; Jerry W. Craig interview, August 12, 1985.

CHAPTER 11: INTERNATIONAL PARTICIPATION

1. Dr. Karl H. Doetsch interview, November 20, 1986. See also NASA, "The Space Operations Center: Review Draws International Participation," *Space News Roundup*, Lyndon B. Johnson Space Center, January 22, 1982, pp. 1–2.

2. See Kenneth S. Pedersen, "The Changing Face of International Space Cooperation: One View of NASA," *Space Policy* 2 (May 1986): 120–137; and John M. Logsdon, "International Involvement in a Civilian Space Station Program" (study prepared under NASA purchase order 3-024-001 to JML, Inc., Laurel, Md., May 1984).

3. Doetsch interview, November 20, 1986.

4. "Snow Prompts Closings, Traffic Woes," *Houston Post*, Thursday, January 14, 1982, pp. A1, A21; Kenneth S. Pedersen interview, October 29, 1986.

5. See Pedersen, "The Changing Face of International Space Cooperation," pp. 123–128; and Marcia S. Smith, *Space Activities of the United States, Soviet Union, and Other Launching Countries/Organizations: 1957–1983* (Washington, D.C.: Congressional Research Service, Library of Congress, January 15, 1984), pp. 127–136.

6. Public Law 85-568, National Aeronautics and Space Act of 1958, 85th Cong., 2d sess., July 29, 1958, sec. 102 (c) (7). See also George M. Low, Memorandum for the Record, Subject: Meeting with the President on January 5, 1972,

dated January 12, 1972, p. 2; and T. O. Paine, NASA Administrator, to the President, November 7, 1969; both in NASA History Office.

7. T. O. Paine to the President, March 26, 1970, p. 2; T. O. Paine to the President, February 12, 1969, p. 2; both in NASA History Office.

8. Logsdon, "International Involvement in a Civilian Space Station Program," pp. 14–17. See also William A. Shumann, "NASA Shuttle Projection Sees Heavy Spacelab Use," *Aviation Week & Space Technology,* November 5, 1973, p. 22.

9. Wolfgang Finke, "Remarks on German Space Policy, 1985 to 1995," in H. Stoewer and Peter M. Bainum, *From Spacelab to Spacestation,* Advances in the Astronautical Sciences, vol. 56 (San Diego: American Astronautical Society Publications, 1985), p. 13.

10. See the collection of articles under the heading "Internationalizing the Space Station," *Commercial Space,* Winter 1986; and Bruno Gire, "The Ariane Launcher," *Spaceflight* 24 (October 1982): 358–364.

11. Doetsch interview, November 20, 1986.

12. Robert F. Freitag, Briefing for the Honorable Harold C. Hollenbeck, July 27, 1982, viewgraph page entitled "International Interests," NASA History Office.

13. Pedersen interview, October 29, 1986; see also Kenneth S. Pedersen, Memorandum to Philip Culbertson, June 30, 1982, NASA History Office.

14. Pedersen interview, October 29, 1986.

15. NASA Office of Space Transportation Systems, Advanced Programs, "Proceedings of Space Station Planning Workshop Held at the NASA/Michoud Assembly Facility, New Orleans, Louisiana, November 18–20, 1981," NASA History Office.

16. Pedersen interview, October 29, 1986; Pedersen, "The Changing Face of International Space Cooperation," p. 129.

17. Pedersen interview, October 29, 1986.

18. Robert F. Freitag interview conducted by Sylvia Fries, July 12, 1985, NASA History Office.

19. NASA, "Proceedings of Space Station Planning Workshop, November 18–20, 1981."

20. Doetsch interview, November 20, 1986.

21. European Space Agency, *ESA News Release,* "ESA Awards Contracts for European Participation in the US Space Station Programme," November 5, 1982, Information no. 35; "ESA to Study Possible Participation in U.S. Space Station," *Defense Daily,* June 24, 1982, p. 303; and Lyn D. Wigbels, Memorandum to John Hodge, Subject: Background for European Trip, May 23, 1983; all in NASA History Office.

22. "Canada Contracts with Spar for Study of Space Station Role," *Defense Daily,* January 12, 1983, p. 55; Doetsch interview, November 20, 1986.

CHAPTER 12: TECHNOLOGY

1. Daniel H. Herman interview, July 19, 1985.

2. Richard Carlisle interview, July 12, 1985.

3. See Nicholas L. Johnson, *The Soviet Year in Space: 1986* (Colorado Springs: Teledyne Brown Engineering, 1987), pp. 53–65.

4. Space Station Technology Steering Committee, Organization Meeting, December 10, 1981, from notes taken by the recording secretary, pp. 3–4, NASA History Office.

5. Daniel H. Herman interview conducted by Sylvia D. Fries, March 26, 1985, NASA History Office.

6. Herman interview, March 26, 1985.

7. Paul Holloway, "Space Station Technology" (paper presented at the Thirty-third International Astronautical Federation [IAF] Congress, Paris, France, September 16, 1982–October 2, 1982, IAF-82-15), p. 2, NASA History Office (published under the same title in the *British Interplanetary Society Journal* 36 [September 1983]: 409–425). See also Richard F. Carlisle, "Status of Planning for a Space Station Technology Steering Group," October 2, 8, 16, 1981, viewgraph labeled "Space Station Technology Steering Committee (Objectives)"; and Space Station Technology Steering Committee, Organization Meeting, December 10, 1981; both in NASA History Office.

8. RSS-5/Manager, Spacecraft Systems Office, [Richard Carlisle], Subject: Space Station Technology Steering Committee, [Minutes of December 10, 1981, Meeting], distributed January 11, 1982, p. 1, NASA History Office.

9. Ibid.; see also RSS-5/Manager, Spacecraft Systems Office, Subject: Top Ten Technology Challenges, January 15, 1982, NASA History Office.

10. RSS-5/Manager, Spacecraft Systems Office, Subject: Space Station Technology Steering Committee Minutes, January 20–21, 1982, at KSC, dated January 27, 1982, NASA History Office.

11. NASA, "Space Station Program Description Document," bk. 4, "Technology and Advanced Development," 1st ed., February 1, 1983, p. 17, NASA History Office.

12. Holloway, "Space Station Technology," pp. 7, 8.

13. Ibid., p. 9.

14. See RSS-5/Manager, Spacecraft Systems Office, Space Station Technology Steering Committee Minutes, January 20–21, 1982, enclosure 4; RSS-5/Manager, Space Station Systems Office, Subject: Minutes of Meeting, June 23–25, 1982, GSFC, distributed July 22, 1982, p. 3; and NASA, "Space Station Program Description Document," bk. 4, 1st ed., p. 10; all in NASA History Office.

15. RSS-5/Executive Secretary, Space Station Technology Steering Committee, Subject: Minutes of Space Station Technology Steering Committee and Working Group Meetings at MSFC, March 15–18, 1982, dated March 25, 1982; NASA, "Space Station Program Description Document," bk. 4, 1st ed.; both in NASA History Office.

16. Herman interview, March 26, 1985; W. David Compton and Charles D. Benson, *Living and Working in Space: A History of Skylab*, SP-4208 (Washington, D.C.: NASA, 1983); NASA Technical Memorandum (TM X-64814), *Skylab: MSFC Skylab Mission Report—Saturn Workshop* (Huntsville, Ala.: George C. Marshall Space Flight Center, October, 1974); Leland Belew and Ernest Stuhlinger, *Skylab: A Guidebook* (Huntsville, Ala.: George C. Marshall Space Flight Center, n.d.).

17. Herman interview, March 26, 1985.

18. NASA Office of Public Affairs, *Skylab: News Reference* (Washington, D.C.: NASA, March 1973).

19. See George Drake, "Crew and Life Support: ECLSS," NASA, *Space Station Technology, 1983* (proceedings of the Space Station Technology Workshop held in Williamsburg, Va., March 28–31, 1983), NASA Conference Publication 2293 (Washington, D.C.: NASA, 1984), p. 36; see also NASA, *MSFC Skylab Mission Report*, pp. 10–13.

20. NASA, "Space Station Program Description Document," bk. 4, 1st ed., p. 13.

21. Ibid., pp. 13–14.

22. Holloway, "Space Station Technology," p. 11.

23. Ibid., p. 11. See also NASA, "Space Station Program Description Document," bk. 4, 1st ed., p. 14.

24. Holloway, "Space Station Technology," p. 6; Allen J. Louviere interview, August 13, 1985.

25. NASA, "Space Station Program Description Document," bk. 4, 1st ed., p. 8; Holloway, "Space Station Technology," p. 16.

26. NASA, Space Station Task Force, "Space Station Program Description Document," bk. 4, "Advanced Development Program," final ed., March 1984, chap. 4, pp. 1–10; NASA, "Space Station Program Description Document," bk. 4, 1st ed., pp. 3–9; both in NASA History Office.

27. John D. Hodge and Richard Carlisle interview conducted by Sylvia Fries, March 4, 1985, NASA History Office.

28. Chairman, Space Station Technology Steering Committee [Walter Olstad], Subject: Thoughts from the Recent Space Station Technology Steering Committee Meeting, May 7, 1982, p. 2, NASA History Office; Holloway, "Space Station Technology," p. 15; NASA, "Space Station Program Description Document," bk. 4, 1st ed., pp. 6, 12.

29. NASA, "Space Station Program Description Document," bk. 4, 1st ed., p. 12.

30. Herman interview, July 19, 1985.

31. Hodge and Carlisle interview, March 4, 1985; Herman interview, March 26, 1985.

CHAPTER 13: BUDGET WARS

1. See RSS-5/Manager, Space Station Systems Office, [Richard Carlisle], Subject: Minutes of Meeting, August 12, 1982, at NASA Headquarters, dated September 17, 1982; and Philip E. Culbertson, Note for Mr. Richard Malow, Subcommittee on HUD and Independent Agencies, Committee on Appropriations, Subject: NASA FY 83 Space Station Budget Activity, August 3, 1982; both in NASA History Office.

2. John Hodge, Space Station Planning Presentation to OMB, September 27, 1982, NASA History Office.

3. NASA Office of Space Station, Business Management Division, [John P. Sheahan, Acting Director], "FY 1984 Budget," vol. 1. See also John P. Sheahan, "Space Station Program Review Committee, FY 1983 Operating Plan and FY 1984 Budget Status," January 26, 1983, especially p. 4 ("Space Station FY 1984 Budget, Augmentation Funding Chronology"); and Hans Mark, "NASA FY 1984 Budget,"

briefing to NASA employees dated February 2, 1983; both in NASA History Office.

4. James M. Beggs to David A. Stockman, Director, OMB, September 15, 1982, p. 3, NASA History Office; Administration of Ronald Reagan, "United States Space Policy: Fact Sheet Outlining the Policy," *Weekly Compilation of Presidential Documents*, July 4, 1982, p. 874; James M. Beggs interview, September 3, 1982.

5. See "Space Station Definition: Funding Status," December 14, 1982, in NASA Office of Space Station, "FY 1984 Budget," vol. 1.

6. John Hodge, Space Station Planning Presentation to OMB, September 27, 1982.

7. U.S. Executive Office of the President, Office of Management and Budget, *Budget of the U.S. Government, Fiscal Year 1984* (Washington, D.C.: Government Printing Office, 1963), appendix, sec. 2, p. 10.

8. "Space Station Definition: Funding Status," December 14, 1982.

9. See the viewgraph labeled "Space Station Definition, FY 1984 Budget OMB Rationale, OMB Claims NASA Is Spending $50–$70 Million a Year on Space Station as Follows," and the notes that follow, labeled "Khedouri Phone Conversation with Beggs/Mark," in NASA Office of Space Station, "FY 1984 Budget," vol. 1. See also John T. Sheahan interview, August 1, 1985; and Beggs interview, September 3, 1985.

10. See "$53M in FY 84 Would Enable," n.d.; and "$53M in FY 84 Would Enable," December 8, 1982; both in NASA Office of Space Station, "FY 1984 Budget," vol. 1. See also Daniel Herman interview, July 29, 1986.

11. "Space Station Definition: Funding Status," December 14, 1982.

12. Beggs to Stockman, September 15, 1982, last page.

13. See STS-105 file, NASA History Office; and U.S. House, Committee on Science and Technology, Subcommittee on Space Science and Applications, *The Need for a Fifth Space Shuttle Orbiter*, 97th Cong., 2d sess., June 15, 1982, pp. 46, 61–62.

14. Sheahan interview, August 1, 1985; see also notes prepared by Sheahan labeled "Conversation with Campbell, 12/15/82, Rationale for Space Station Mark," in NASA Office of Space Station, "FY 1984 Budget," vol. 1.

15. Claiborne R. Hicks interview, July 17, 1985.

16. Beggs interview, September 3, 1985.

17. Jack Young interview, January 16, 1986.

18. Hans Mark, *The Space Station: A Personal Journey* (Durham, N.C.: Duke University Press, 1987), p. 153; "Schmitt: White House Balks at New Space Initiatives," *Aerospace Daily*, August 24, 1982, p. 301. See also Lee D. Saegesser, Memo for the Record, December 2, 1982, NASA History Office.

19. David Stockman, *The Triumph of Politics* (New York: Harper & Row, 1986), p. 150.

20. Beggs interview, September 3, 1985.

21. Dr. Hans Mark, Charts Used by Dr. Mark in Presentation to NASA Employees, February 2, 1983, p. 1, NASA History Office.

22. James Beggs, Fiscal Year 1984 Budget Briefing with Mr. Beggs before NASA, January 31, 1983, p. 20, NASA History Office.

23. Beggs interview, September 3, 1985.

CHAPTER 14: POSITIONS

1. David Stockman, *The Triumph of Politics* (New York: Harper and Row, 1986), fig. 43.

2. Ibid., especially p. 21; "David Stockman," *Current Biography* (New York: H. W. Wilson Co., 1981), pp. 400–403; Walter Shapiro, "The Stockman Express," *Washington Post Magazine,* February 8, 1981, p. 11.

3. David A. Stockman, "The Social Pork Barrel," *The Public Interest* 39 (Spring 1975): 3–30.

4. U.S. Senate, Committee on the Budget, *First Concurrent Resolution on the Budget—Fiscal Year 1986,* 99th Cong., 1st sess., 1985, vol. 2, p. 61.

5. Stockman, *The Triumph of Politics,* p. 151; "Stockman and the 'One-third' Cutback in the NASA Budget," *Defense Daily,* January 6, 1981, pp. 6–7.

6. U.S. Senate, Appropriations Committee, a Subcommittee, *Department of Housing and Urban Development, and Certain Independent Agencies Appropriations, Fiscal Year 1983,* 97th Cong., 2d sess., 1982, pt. 2, pp. 1089, 1088.

7. U.S. Executive Office of the President, *America's New Beginning: A Program for Economic Recovery* (Washington, D.C.: Executive Office of the President, February 18, 1981), attachment labeled "Reductions in National Aeronautics and Space Administration Programs," sec. 6, pp. 35–36, NASA History Office; Stockman, *The Triumph of Politics,* p. 151.

8. George Keyworth interview conducted by Shirley H. Scheibla, "Magic Formula?" *Barron's* 62 (December 6, 1982): 9.

9. U.S. House, Committee on Appropriations, Subcommittee on HUD-Independent Agencies, *Department of Housing and Urban Development–Independent Agencies Appropriations for 1984,* 98th Cong., 1st sess., 1983, pt. 3, p. 19.

10. J. A. Van Allen, *New York Times,* April 1, 1986, p. A31; James A. Van Allen, "Space Science, Space Technology, and the Space Station," *Scientific American* 254 (January 1986): 36.

11. U.S. Senate, Committee on Appropriations, a Subcommittee, *Department of Housing and Urban Development and Certain Independent Agencies Appropriations for Fiscal Year 1984,* 98th Cong., 1st sess., 1983, pt. 1, p. 553.

12. Jerome Wiesner interview conducted by John Logsdon, quoted in Logsdon, *The Decision to Go to the Moon* (Cambridge, Mass.: MIT Press, 1970), p. 118; David Bell interview conducted by John Logsdon, October 4, 1967, NASA History Office.

13. Philip J. Hilts, "Physicist, Relatively Unknown, Named as Reagan's Science Adviser," *Washington Post,* June 14, 1981, p. A12.

14. Quoted from Lee Walczak, "Washington Outlook: How a Supply-sider Would Run OMB," *Business Week,* December 15, 1980, p. 125.

15. Logsdon, *The Decision to Go to the Moon,* p. 123.

16. Robert C. Seamans, Jr., Secretary of the Air Force, to Spiro T. Agnew, Vice President of the United States, August 4, 1969, p. 4; Thomas O. Paine, Administrator, NASA, and Robert C. Seamans, Secretary of the Air Force, "Agreement Between the National Aeronautics and Space Administration and the Department of the Air Force Concerning the Space Transportation System," February 17, 1970, p. 1; James C. Fletcher, Administrator, NASA, and William P. Clements, Deputy Secretary of Defense, "NASA/DOD Memorandum of Understanding on Management and Oper-

ation of the Space Transportation System," January 14, 1977, p. 8; A. M. Lovelace, Deputy Administrator, NASA, and Robert N. Parker, Acting Director, Defense Research and Engineering, DOD, "Memorandum of Agreement Between NASA and DOD: Basic Principles for NASA/DOD Space Transportation System Launch Reimbursement," March 7, 1977, p. 1; Lynn Heninger, Legislative Affairs Specialist, memorandum to Wally Berger, May 21, 1982; all in NASA History Office files.

17. Hans Mark, *The Space Station: A Personal Journey* (Durham, N.C.: Duke University Press, 1987), pp. 69–70, 99, 106–108; Hans Mark interview, November 11, 1987.

18. See "DOD Said to Have Cut MOL Without Air Force Advice," *Aerospace Daily,* June 13, 1967, pp. F5–F6; and NASA History Office, MOL program files.

19. Quoted from Charles W. Corddry, "Orbital Lab Scrapped by Pentagon," *Baltimore Sun,* June 11, 1969; headlines from *Space Daily,* September 16, 1969, p. 199, and October 19, 1972, p. 222.

20. John Hodge interview, September 16, 1985, pp. 1–2.

21. William E. Burrows, *Deep Black: Space Espionage and National Security* (New York: Random House, 1987).

22. Seamans to Agnew, August 4, 1969, p. 4.

23. Quoted from "Air Force Leader Says Manned Military Space Stations Not Needed," *Defense/Space Business Daily,* February 16, 1979, p. 243.

24. Quoted from "Weinberger Sees Solar Power Satellites/Space Stations," *Defense Daily,* May 22, 1981, p. 124; U.S. Senate, Committee on Commerce, Science, and Transportation, Subcommittee on Science, Technology, and Space, *NASA Authorization for Fiscal Year 1984,* 98th Cong., 1st sess., 1983, p. 17.

25. James Beggs interview, September 3, 1985; John Hodge, "Presentation on Space Station Planning to the NASA Center Directors," September 22, 1982, NASA History Office; Mark, *Space Station,* pp. 162, 178.

26. Mark, *Space Station,* p. 171; "NASA/DOD Memorandum of Understanding," January 14, 1977, p. 8.

27. Mark, *Space Station,* pp. 93, 97–99, 172, 224.

28. Senate Subcommittee on Science, Technology, and Space, *NASA Authorization for Fiscal Year 1984,* p. 17.

CHAPTER 15: THE WHITE HOUSE

1. See Richard F. Fenno, *The President's Cabinet* (New York: Vintage Books, 1959), p. 29; and Thomas E. Cronin, *The State of the Presidency,* 2d ed. (Boston: Little, Brown & Co., 1980), p. 11.

2. Allan Nevins, *The War for the Union,* vol. 1 (New York: Charles Scribner's Sons, 1959), pp. 45–46, 55–56.

3. U.S. Office of the Federal Register, *U.S. Government Manual, 1984/85* (Washington, D.C.: Government Printing Office, 1984), p. 77.

4. See William Ryan and Desmond Guinness, *The White House: An Architectural History* (New York: McGraw-Hill Book Co., 1980); and Howard E. McCurdy, "Crowding and Behavior in the White House," *Psychology Today* 15 (April 1981): 21–25.

5. James M. Beggs interview, October 30, 1987. See also Gilbert Rye interview, January 15, 1988.

6. James Beggs interview, September 3, 1985.

7. Beggs interview, October 30, 1987.

8. The White House, National Security Study Directive 5-83, "Space Station," April 11, 1983, NASA History Office (hereafter cited as NSSD 5-83).

9. Administration of Ronald Reagan, "United States Space Policy: Fact Sheet Outlining the Policy," *Weekly Compilation of Presidential Documents*, July 4, 1982, p. 875.

10. See R. Jeffrey Smith, "Squabbling Over the Space Policy," *Science* 217 (July 23, 1982): 333; Gilbert D. Rye interview, September 13, 1985; and Philip Culbertson interview, September 14, 1987.

11. Rye interview, January 15, 1988; see also Hans Mark interview, November 11, 1987.

12. NASA, "Proceedings of Space Station Planning Workshop Held at the NASA/Michoud Assembly Facility in New Orleans, Louisiana, November 18–20, 1981," NASA History Office.

13. Gilbert Rye to Hans Mark, October 25, 1984.

14. Rye interview, September 13, 1985.

15. "United States Space Policy: Fact Sheet," July 4, 1982, p. 875.

16. NSSD 5-83, p. 2; see also William P. Clark, Action Memorandum for the President, Subject: Space Station, April 7, 1983.

17. Rye interview, September 13, 1985.

18. NSSD 5-83, p. 2.

19. Beggs interview, September 3, 1985.

CHAPTER 16: THE RABBIT IN THE HAT

1. U.S. House, Committee on Science and Technology, Subcommittee on Space Science and Applications, *NASA's Space Station Activities*, 98th Cong., 1st sess., 1983, pp. 3–35; see also Space Station Task Force briefing files, NASA History Office.

2. Administration of Ronald Reagan, "Remarks on the Completion of the Fourth Mission of the Space Shuttle Columbia," *Weekly Compilation of Presidential Documents*, July 4, 1982, p. 870.

3. House Committee on Science and Technology, *NASA's Space Station Activities*, pp. 3–6, 15–21.

4. Ibid., p. 8; see also NASA Space Station Task Force, "Space Station Needs, Attributes, and Architectural Options (Mission Analysis Studies)," Document Control List (5 pp.), NASA History Office.

5. Boeing Aerospace Company, "Space Station Needs, Attributes, and Architectural Options Study," vol. 6, "Final Report, Final Briefing," contract NASW-3680, NASA Headquarters, Washington, D.C., April 5, 1983, p. 21; Lockheed Missiles and Space Company, "NASA Space Station Needs, Attributes, and Architectural Options," vol. 1, "Executive Summary NASA, Final Presentation," contract NASW-

3684, NASA Headquarters, Washington, D.C., p. 8. See also Brian Pritchard interview, August 23, 1985.

6. House Committee on Science and Technology, *NASA's Space Station Activities*, p. 11.

7. The Langley Data Base provided a computerized repository of data on the missions associated with the space station. See Pritchard interview, August 23, 1985, p. 8.

8. NASA, Space Station Task Force, "Space Station Program Description Document," bk. 2, "Mission Description Document," March 1984, chap. 3, p. 27, chap. 6, p. 3; see also Pritchard interview, August 23, 1985.

9. NASA, Space Station Task Force, "Space Station Program Description Document," bk. 2, chap. 3, p. 8, chap. 6, p. 2.

10. Ibid., bk. 2, chap. 3, p. 35, chap. 6, p. 6; McDonnell Douglas Astronautics Company, "Space Station Needs, Attributes, and Architectural Options," Final Study Report, Summary Briefing, April 1983, NASA Headquarters, Washington, D.C., pp. 14, 58, 59; Pritchard interview, August 23, 1985.

11. NASA, Space Station Task Force, "Space Station Program Description Document," bk. 2, chap. 3, pp. 80–81, chap. 6, pp. 10–11.

12. Ibid., bk. 2, chap. 3, p. 101, chap. 5, p. 3, chap. 6, p. 13.

13. Ibid., bk. 2, chap. 1, pp. 3–4; Lockheed, "Space Station Needs," vol. 1, p. 8.

14. See House Committee on Science and Technology, *NASA's Space Station Activities*, pp. 4–6; and NASA, Space Station Task Force, "Space Station Program Description Document," bk. 2, chap. 3, p. 101.

15. Paul B. Stares, *The Militarization of Space* (Ithaca, N.Y.: Cornell University Press, 1985).

16. See "A.F. Eyes Space Station 'Shopping List,' Awaits Shuttle Experience," *Aerospace Daily*, April 27, 1982, pp. 329–330; and "Department of Defense Requirements Review for SIG (Space) Manned Space Station Study," attachment to Memorandum from Charles W. Cook to General Stilwell, Subject: DOD Space Station Requirements, June 3, 1983, NASA History Office.

17. Pritchard interview, August 23, 1985.

18. "Department of Defense Requirements Review for SIG (Space)," p. 3; and Memorandum from Charles W. Cook to General Stilwell, June 3, 1983, p. 1. See also Richard G. Stilwell, General, U.S.A. (Ret.), to Robert C. McFarlane, Deputy Assistant to the President for National Security Affairs, June 20, 1983, NASA History Office.

19. Pritchard interview, August 23, 1985.

20. NASA, Space Station Task Force, "Space Station Program Description Document," bk. 2, chap. 1, p. 4.

21. See, for example, TRW, "Space Station Needs, Attributes, and Architectural Options Study," Final Review Executive Summary Briefing, contract NASW-3681, NASA Headquarters, Washington, D.C., April 5, 1983, p. 21; and Boeing, "Space Station Needs, Attributes, and Architectural Options Study," vol. 6, pp. 31–34.

22. NASA, NSM-23/GAO Liaison Officer to Distribution, Subject: GAO Draft Letter Report Entitled "NASA Has Prematurely Focused Its Planning on a Space Station as Its Next Major Program (Code 951709)," January 18, 1983, with attach-

ment; W. H. Sheley, Director, U.S. General Accounting Office, to James M. Beggs, Administrator, NASA, Subject: NASA Needs to Broaden Its Planning Before Selecting Its Next Major Space Program (GAO/MASAD-83-23), p. 4; all in NASA History Office.

23. NASA, GAO Draft Letter, pp. 6–7; Sheley to Beggs, p. 4.

24. Philip E. Culbertson, Associate Deputy Administrator, NASA, "NASA Comments on GAO Draft Report on Space Station Planning (a GAO review under Code 951709)," undated but issued approximately February 1983, p. 2, NASA History Office.

25. House Committee on Science and Technology, *NASA's Space Station Activities*, testimony of Kenneth S. Pedersen, p. 93.

26. Cecil Gregg interview, November 19, 1985; Clarke Covington interview, August 12, 1985.

27. Luther Powell interview, November 18, 1985.

28. William Marshall interview, November 18, 1985; James Fletcher interview, October 14, 1985; "Space Station: Planning Approach," Space Station Task Force briefing file labeled "John's pitch at [NASA's] Williamsburg [Technology Conference], March 28, 1983."

29. Powell interview, November 18, 1985.

30. Ibid.; House Committee on Science and Technology, *NASA's Space Station Activities*, p. 11.

31. House Committee on Science and Technology, *NASA's Space Station Activities*, p. 65; U.S. Senate, Committee on Commerce, Science, and Transportation, Subcommittee on Science, Technology, and Space, *Civil Space Station*, 98th Cong., 1st sess., November 15, 1983, p. 55.

32. Powell interview, November 18, 1985; NASA, Space Station Task Force, "Space Station Program Description Document," bk. 2, chap. 5, p. 3; NASA, Space Station Task Force, "Space Station Program Description Document," bk. 3, "System Requirements and Characteristics," March 1984, chap. 2, pp. 2–3, NASA History Office.

33. Powell interview, November 18, 1985.

34. House Committee on Science and Technology, *NASA's Space Station Activities*, pp. 11–12, 28, 31; Senate Committee on Commerce, Science, and Transportation, *Civil Space Station*, pp. 49, 54–55; NASA, Space Station Task Force, "Space Station Program Description Document," bk. 1, "Introduction and Summary," March 1984, chap. 5, pp. 1–4, NASA History Office.

35. Powell interview, November 18, 1985.

36. House Committee on Science and Technology, *NASA's Space Station Activities*, pp. 13–14.

CHAPTER 17: SIG (Space)

1. John Hodge interview, September 16, 1985; see also NASA Space Station Task Force, "Space Station Management Colloquium White Paper," August 1983, p. 2, NASA History Office.

2. Daniel Herman interview conducted by Sylvia D. Fries, March 26, 1985; see also Philip Culbertson interview, December 11, 1987.

3. See, for example, U.S. Senate, Committee on Commerce, Science, and Transportation, Subcommittee on Science, Technology, and Space, *Civil Space Station,* 98th Cong., 1st sess., 1983, testimony of Thomas M. Donahue, Chairman, Space Science Board, National Academy of Sciences, pp. 61–64.

4. Thomas M. Donahue, Chairman, Space Science Board, National Research Council, National Academy of Sciences, to James Beggs, NASA Administrator, September 9, 1983, with attachments, reprinted in Senate Committee on Commerce, Science, and Transportation, *Civil Space Station,* pp. 64–66.

5. Senate Committee on Commerce, Science, and Transportation, *Civil Space Station,* pp. 20–21.

6. The White House, National Security Study Directive 5-83, "Space Station," April 11, 1983, p. 1, NASA History Office (hereafter cited as NSSD 5-83).

7. "Argumentation for Space Station Option (from SIG(Space) Issue Paper)," attachment to James M. Beggs to the Honorable James A. Baker, August 24, 1983, NASA History Office.

8. "A NASA Capabilities Evaluation Document [SIG Document]," Preliminary Draft, June 24, 1983, NASA History Office.

9. Testimony of Thomas Donahue, Senate Committee on Commerce, Science, and Transportation, *Civil Space Station,* p. 62.

10. "Space Science Board Assessment of the Scientific Value of a Space Station," attachment to Donahue letter to Beggs, September 9, 1983, reprinted in Senate Committee on Commerce, Science, and Transportation, *Civil Space Station,* p. 65.

11. George A. Keyworth, "The U.S. Space Program—Where Do We Go From Here?" (proposed remarks to the Nineteenth Joint Propulsion Conference, Seattle, Washington, June 27, 1983), pp. 1, 11, 13, 14, NASA History Office.

12. Ronald Reagan, "National Security: Address to the Nation," *Weekly Compilation of Presidential Documents,* March 23, 1983, pp. 447–448.

13. "DeLauer Sees 10-Year/$30–$40 Billion SDI Effort," *Defense Daily,* May 4, 1984, p. 27; "Ten Year Cost of SDI Put at $670-$770 Billion," *Defense Daily,* July 23, 1986, p. 122; Administration of Ronald Reagan, "U.S. Space Policy: Fact Sheet," *Weekly Compilation of Presidential Documents,* July 4, 1982, p. 874; Ivan Bekey interview, January 29, 1988. See also U.S. National Commission on Space, *Pioneering the Space Frontier* (New York: Bantam Books, 1986), especially fig. 2 on p. 16.

14. Brian Pritchard interview, August 23, 1985.

15. Hodge interview, September 16, 1985; see also the viewgraph labeled "Space Station Planning: 'The SIG,'" John D. Hodge Briefing for the Honorable Manuel Lujan, Subcommittee on Space Science and Applications, House of Representatives, February 18, 1983, NASA History Office.

16. See Margaret Finarelli interview, August 20, 1985.

17. Gilbert Rye interview, January 15, 1988; Information Memorandum for William P. Clark from Gilbert D. Rye, Subject: Manned Space Station, March 17, 1983.

18. NSSD 5-83, p. 2 (emphasis added).

19. Finarelli interview, August 20, 1985.

20. Rye interview, January 15, 1988; Finarelli interview, August 20, 1985.

21. See Malcolm Baldrige, Secretary of Commerce, to the Honorable William P. Clark, Assistant to the President for National Security Affairs, August 23, 1983, NASA History Office.

22. "Argumentation for Space Station Option," attachment to Beggs to Baker, August 24, 1983.

23. Senior Interagency Group for Space, "Space Station Report," prepared by NSC drafting team, Gilbert D. Rye, Chairman, August 4, 1983.

24. See Charles Cook interview, December 9, 1986.

25. Hans Mark, *The Space Station: A Personal Journey* (Durham, N.C.: Duke University Press, 1987), pp. 176–177.

26. James Beggs interview, September 3, 1985.

27. "Department of Defense Requirements Review for SIG (Space), Manned Space Station Study," attachment to Memorandum from Charles W. Cook, Department of the Air Force, to General Stilwell, Subject: DOD Space Station Requirements, June 3, 1983, p. 3, NASA History Office. See also U.S. Air Force, Scientific Advisory Board, "Report of the USAF Scientific Advisory Board Ad Hoc Committee on the Potential Military Utility of a Manned National Space Station," June 1983, Air University Library, Maxwell AFB, Alabama; and James M. Beggs, NASA Administrator, to Caspar W. Weinberger, Secretary of Defense, August 19, 1983, NASA History Office.

28. Caspar Weinberger, Secretary of Defense, to James M. Beggs, NASA Administrator, January 16, 1984, NASA History Office.

29. "Department of Defense Requirements Review for SIG (Space)," attachment to Memorandum from Cook to Stilwell, pp. 3–4.

30. Weinberger to Beggs, January 16, 1984.

31. Beggs to Weinberger, August 19, 1983.

32. "Argumentation against Deferral Option (from SIG(Space) Issue Paper), attachment to Beggs to Baker, August 24, 1983.

33. Beggs interview, September 3, 1985.

34. Weinberger to Beggs, January 16, 1984.

35. Beggs interview, September 3, 1985.

CHAPTER 18: THE NUMBER

1. U.S. Senate, Committee on Commerce, Science, and Transportation, Subcommittee on Science, Technology, and Space, *NASA Authorization for Fiscal Year 1984*, 98th Cong., 1st sess., March 9, 1983, testimony of James Beggs, p. 51.

2. See "Some Space Station Budgetary Assumptions," in "Space Station Presentation to the NASA Advisory Council," July 21, 1982; "Space Station Planning: Acquisition," in John Hodge, "Presentation on Space Station Planning to the NASA Center Directors," September 22, 1982; "Space Station Planning Approach," in John Hodge, "Space Station Planning Presentation to OMB," September 27, 1982; all in NASA History Office. See also Luther Powell interview, November 18, 1985.

3. "Space Station Capabilities," in John Hodge, "Space Station Presentation to Dr. Charles W. Cook," May 23, 1983, NASA History Office.

4. Powell interview, November 18, 1985.

5. Philip Culbertson interview, December 11, 1987; John Hodge interview, July 10, 1985.

6. James Beggs interview, October 30, 1987.

7. Edwin Meese interview, January 8, 1986.

8. Culbertson interview, December 11, 1987.

9. See John D. Hodge, "Space Station Presentation to the Space and Earth Science Advisory Committee," October 25, 1983, and other Space Station Task Force briefing files, NASA History Office.

10. U.S. Senate, Committee on Commerce, Science, and Transportation, Subcommittee on Science, Technology, and Space, *Civil Space Station*, 98th Cong., 1st sess., 1983, p. 49.

11. James C. Miller, Director, Office of Management and Budget, Memorandum for the President, February 10, 1987; U.S. Congressional Budget Office, *The NASA Program in the 1990s and Beyond* (Washington, D.C.: Congressional Budget Office, May 1988), p. 25; National Research Council, *Report of the Committee on the Space Station* (Washington, D.C.: National Academy Press, September 1987), pp. 30–31.

12. Peggy Finarelli to OMB/Bart Borrasca, Subject: Space Station Funding, September 8, 1983, NASA History Office; see also Senate Committee on Commerce, Science, and Transportation, *Civil Space Station*, p. 55.

13. Miller Memorandum for the President, February 10, 1987; John Hodge to Sylvia Fries, June 14, 1989, NASA History Office.

14. Miller, Memorandum for the President, February 10, 1987.

15. Space Station Task Force roundtable interview, August 4, 1987; Culbertson interview, December 11, 1987.

16. Miller, Memorandum for the President, February 10, 1987; NASA, "NASA Proceeding Toward Space Station Development," *NASA News*, release no. 87-50, April 3, 1987.

17. Space Station Task Force roundtable interview, August 4, 1987.

18. John Hodge, Memorandum to Dale Myers, November 3, 1986, NASA History Office; Miller, Memorandum for the President, February 10, 1987; National Research Council, *Report of the Committee on the Space Station*, p. 30.

19. Congressional Budget Office, *The NASA Program in the 1990s and Beyond*, pp. 26–27; National Research Council, *Report of the Committee on the Space Station*, pp. 30–31.

20. NASA Office of Space Station, "Space Station Capital Development Plan, Fiscal Year 1989," April 1988, p. 4, NASA History Office; NASA Office of Space Station, *The Space Station: A Description of the Configuration Established at the Systems Requirements Review (SSR)* (Washington, D.C.: Technical and Administrative Services Corporation, June 1986), p. 16.

21. Hodge, Memorandum to Dale Myers, November 3, 1986; John Hodge interview, July 10, 1985.

22. Senior Interagency Group for Space, "Space Station Report," prepared by NSC drafting team, Gilbert D. Rye, chairman, August 4, 1983; Hodge interview, July 10, 1985.

23. Hodge interview, July 10, 1985; James M. Beggs to the Honorable James A. Baker, August 24, 1983; Peggy Finarelli to OMB/Bart Borrasca, September 8, 1983.

24. Powell interview, November 18, 1985.

25. U.S. House, Committee on Science and Technology, Subcommittee on Space Science and Applications, *NASA's Space Station Activities*, 98th Cong., 1st sess., 1983, pp. 7–8, 23; Hodge interview, July 10, 1985; Philip Culbertson interview, May 23, 1989; Margaret Finarelli interview, February 9, 1990; Norman Terrell interview, February 9, 1990.

26. Powell interview, November 18, 1985.

27. Compare House Committee on Science and Technology, *NASA's Space Station Activities*, p. 31, and Finarelli to OMB/Borrasca, September 8, 1983, to Senate Committee on Commerce, Science, and Transportation, *Civil Space Station*, pp. 49, 54–55.

CHAPTER 19: REAGAN

1. "Excerpts of Remarks by Governor Ronald Reagan," Republican dinner, San Diego, California, July 23, 1971, p. 1, Hoover Institution on War, Revolution, and Peace, Stanford, California.

2. "Excerpts of Remarks by Governor Ronald Reagan," American Legion State Convention, Los Angeles, California, June 25, 1971, pp. 2–4, Hoover Institution on War, Revolution, and Peace, Stanford, California.

3. "Excerpts of Remarks by Governor Ronald Reagan," American Legion State Convention, Sacramento, California, June 26, 1970, p. 4, Hoover Institution on War, Revolution, and Peace, Stanford, California.

4. See "Reagan on Aerospace, Aviation," *Aerospace Daily*, September 30, 1980, pp. 167–168.

5. Edwin Meese interview, January 8, 1987. See also "Excerpts of Remarks by Governor Ronald Reagan," Host Breakfast, Sacramento, California, September 4, 1970, p. 3, Hoover Institution on War, Revolution, and Peace, Stanford, California; and Harry Farrell, "Perfect Squelch," *San Jose Mercury*, July 28, 1970.

6. Meese interview, January 8, 1987.

7. Ibid.

8. Michiel Schwarz and Paul Stares, eds., *The Exploitation of Space: Policy Trends in the Military and Commercial Uses of Outer Space* (London: Butterworth & Co., 1985).

9. See Craig Covault, "Reagan Briefed on Space Station," *Aviation Week & Space Technology*, August 8, 1983, pp. 16–18; and Gilbert Rye to Hans Mark, October 25, 1984, p. 3.

10. Covault, "Reagan Briefed on Space Station," p. 16; and Margaret Finarelli interview, August 20, 1985.

11. Rye to Mark, October 25, 1984; and Gilbert Rye interview, September 13, 1985.

12. Ronald Reagan, "National Aeronautics and Space Administration, Remarks at the 25th Anniversary Celebration," *Weekly Compilation of Presidential Documents*, October 19, 1983, pp. 1462–1463.

13. Hans Mark, *The Space Station: A Personal Journey* (Durham, N.C.: Duke University Press, 1987), p. 182; John Hodge interview, September 16, 1985; James M. Beggs interview, October 30, 1987.

14. Craig Fuller interview, April 2, 1987.

15. Terence T. Finn interview conducted by Sylvia Fries, June 12, 1985, NASA History Office.

16. James M. Beggs, "Presentation on Space Station," December 1, 1983, NASA History Office.

17. Hodge interview, September 16, 1985; Mark, *Space Station*, pp. 184–187, 253–254.

18. James M. Beggs interview, September 3, 1985.

19. William Lowther, "The New Race to Colonize the Heavens," *Maclean's* 97 (February 6, 1984): 46–47; Hodge interview, September 16, 1983; "Space Station Decision Overrode Strong Opposition," *Aviation Week & Space Technology*, January 30, 1984, p. 16; Mark, *Space Station*, p. 186.

20. Hodge interview, September 16, 1985; Mark, *Space Station*, pp. 186–187.

21. Mark, *Space Station*, pp. 181–182, 190.

22. Rye to Mark, October 25, 1984.

23. Beggs interview, September 3, 1985.

24. Beggs interview, September 3, 1985; Meese interview, January 8, 1986; Memorandum for Robert C. McFarlane from Gilbert D. Rye, Subject: Space Station Decision, December 8, 1983.

CHAPTER 20: CONGRESS

1. NASA, *Current News*, AP and UPI newswires, January 24, 1984, NASA History Office.

2. U.S. Congress, *Congressional Record*, 98th Cong., 2d sess., January 25, 1984, p. 377.

3. Ronald Reagan, "The State of the Union: Address Delivered Before a Joint Session of the Congress," *Weekly Compilation of Presidential Documents*, January 25, 1984, p. 87.

4. Ibid., pp. 87–90.

5. Robert F. Freitag to Howard McCurdy, February 7, 1987, NASA History Office.

6. Reagan, "State of the Union," January 25, 1984, p. 90; Freitag to McCurdy, February 7, 1987.

7. John F. Kennedy, "Freedom's Cause," May 25, 1961, *Vital Speeches of the Day* 23 (June 15, 1961): 519; Theodore C. Sorensen, *Kennedy* (New York: Harper & Row, 1965), p. 526.

8. Reagan, "State of the Union," January 25, 1984, p. 90 (emphasis added).

9. John Logsdon, *The Decision to Go to the Moon* (Cambridge, Mass.: MIT Press, 1970), pp. 36, 128; NASA Office of Administration, Budget Operations Division, "Chronological History, FY 1973 Budget Submission," September 5, 1972; NASA Comptroller, Budget Operations Division, "Chronological History, FY 85 Budget Submission," final version (undated); James E. Miller, Memorandum for the President, Revised Cost Estimates for the Space Station (attachment), February 10, 1987; last three in NASA History Office.

10. U.S. House, Committee on Appropriations, Subcommittee on HUD-Inde-

pendent Agencies, *Department of Housing and Urban Development–Independent Agencies Appropriations for 1985*, 98th Cong., 2d sess., 1984, pt. 6, p. 2.

11. Edward Boland, "Remarks Prepared for Delivery Before the National Space Club, November 20, 1985," p. 8, NASA History Office.

12. Ibid., pp. 5–7.

13. House Committee on Appropriations, *Department of Housing and Urban Development–Independent Agencies Appropriations for 1985*, pt. 4, p. 27.

14. James M. Beggs, NASA Administrator, to Craig L. Fuller, Assistant to the President for Cabinet Affairs, April 12, 1984; see also U.S. Senate, Committee on Appropriations, a Subcommittee, *Department of Housing and Urban Development and Certain Independent Agencies Appropriations for Fiscal Year 1985*, 98th Cong., 2d sess., 1984, pt. 2, p. 1233.

15. Boland, "Remarks," pp. 10, 9.

16. Bill Green, "Manned or Unmanned Flights?" *New York Times*, March 3, 1986, p. A15.

17. Tina Rosenberg, "Mission Out of Control," *The New Republic*, May 14, 1984, pp. 18–21; "Ivory Tower in Space," *Nature* 307 (January 5, 1984): 2; Thomas Donahue as quoted in Sharon Begley, Mary Hager, and John Carey, "The Next Step Into Space," *Newsweek*, February 6, 1984, p. 83.

18. "An Expensive Yawn in Space," *New York Times*, January 29, 1984, sec. 4, p. 18E.

19. U.S. Congress, Office of Technology Assessment, *Civilian Space Stations and the U.S. Future in Space* (Washington, D.C.: Government Printing Office, 1984), p. 4.

20. Senate Committee on Appropriations, *Department of Housing and Urban Development, and Certain Independent Agencies Appropriations for Fiscal Year 1985*, pt. 2, p. 1245; see also Hans Mark, *The Space Station: A Personal Journey* (Durham, N.C.: Duke University Press, 1987), p. 204.

21. Congressional Quarterly, *1974 CQ Almanac* (Washington, D.C.: Congressional Quarterly Service, 1975), pp. 92, 94; Congressional Quarterly, *1975 CQ Almanac* (Washington, D.C.: Congressional Quarterly Service, 1976), pp. 811, 814; NASA Comptroller, Office of Budget Operations, "Chronological History: FY 1976 and Transition Period Budget Submission," November 11, 1975; NASA Comptroller, Budget Operations Division, "Chronological History: FY 1978 Budget Submission," October 19, 1977; Congressional Quarterly, *Weekly Report*, July 23, 1977, p. 1530.

22. U.S. Congress, *Congressional Record*, 95th Cong., 1st sess., July 19, 1977, pp. 23668, 23673.

CHAPTER 21: MOMENTUM

1. Robert C. McFarlane to James M. Beggs, February 25, 1984, NASA History Office; Gil Rye to Hans Mark, October 25, 1984.

2. Aleksey Popov, "NASA 'Demands' Japan Join in 'Star Wars' Plans," Moscow *TASS* in English, March 12, 1984; Aleksey Popov, "Allies 'Eager' to Follow Pentagon to Space," Moscow *TASS* in English, March 16, 1984; both in NASA History Office.

3. James M. Beggs to George P. Shultz, Secretary of State, March 16, 1984. See also "Mitterrand Sees Space Station for Military Use," *Defense Daily,* October 26, 1984, p. 289; and "Nakasone, NASA Chief Discuss Space Development," Tokyo *Kyodo* in English, March 12, 1984. All in NASA History Office.

4. James M. Beggs to Mr. Isurugi, Minister of State for Science and Technology Agency, Tokyo, Japan, April 6, 1984 (a standard letter to the international participants), NASA History Office.

5. Lynette Wigbels interview, May 22, 1987.

6. Dr. Karl H. Doetsch interview, November 20, 1986. See also Canadian Secretary of State for External Affairs, *Satellites: The Canadian Experience* (Ottawa: Government of Canada, 1984), pp. 20–21; and Government of Canada, News Release, "Canadian Space Program," May 12, 1986, Ottawa.

7. Clive Willis and Wally Cherwinski interview, November 20, 1986.

8. Hans Mark, *The Space Station: A Personal Journey* (Durham, N.C.: Duke University Press, 1987), p. 160.

9. Willis and Cherwinski interview, November 20, 1986.

10. Beggs to Shultz, March 16, 1984.

11. Administration of Ronald Reagan, "London Economic Summit, London Economic Declaration," *Weekly Compilation of Presidential Documents,* June 8, 1984, p. 859.

12. See European Space Agency Council, Resolution on the Long-Term European Space Plan, January 31, 1985, European Space Agency, Paris; and Jeffrey M. Lenorovitz, "Europeans Exploring Independent Role in Space," *Commercial Space,* Winter 1986, pp. 25–29.

13. See "Japan Challenging Western Leadership in Space," *Aviation Week & Space Technology,* July 14, 1986, pp. 18–22; "Japan Preparing New Spaceport Facilities for H-1, H-2 Rockets," *Aviation Week & Space Technology,* July 14, 1986, pp. 51–53; and Eiichiro Sekigawa, "Life Sciences Module for Space Station Will Be Made in Japan," *Commercial Space,* Winter 1986, pp. 37–39.

14. See Lenorovitz, "Europeans Exploring Independent Role in Space," p. 29; European Space Agency Council, Resolution on the Long-Term European Space Plan, p. 2; and "Japan Challenging Western Leadership in Space," *Aviation Week and Space Technology,* p. 19.

15. NASA Space Station Task Force, "Space Station Program Description Document," bk. 4, "Advanced Development Program," March 1984, chap. 2, p.1; see also NASA, *Space Station Technology, 1983,* NASA Conference Publication 2293, Proceedings of the Space Station Technology Workshop Held at the National Conference Center, Williamsburg, Virginia, March 28–31, 1983 (Washington, D.C.: NASA, 1984).

16. Beggs to Isurugi, April 6, 1984; NASA, "Space Station Guidelines for International Cooperation" (prepared in the summer or fall of 1984), p. 2, NASA History Office.

17. See David Dickson, "Space Station Plan Upsets Europe," *Science* 234 (December 19, 1986): 1487; see also Bill Green and Edward P. Boland to William R. Graham, January 14, 1986, NASA History Office.

18. European Space Agency, ESA News Release, "Successful Outcome of ESA Council Meeting at Ministerial Level," January 31, 1985, p. 2, European Space

Agency, Paris; Kenneth S. Pedersen interview, October 29, 1986. See also Peter Marsh, "NASA Tries to Sell W. Europe and Japan on Space Station Project," *Christian Science Monitor,* March 12, 1984, p. 17.

19. Beggs to Shultz, March 16, 1984; see also Pedersen interview, October 29, 1986.

20. Pedersen interview, October 29, 1986; Wigbels interview, May 22, 1987.

21. Memorandum of Understanding Between the National Aeronautics and Space Administration and the European Space Agency for the Conduct of Parallel Detailed Definition and Preliminary Design Studies (Phase B) Leading Toward Further Cooperation in the Development, Operation, and Utilization of a Permanently Manned Space Station, Paris, June 3, 1985; Memorandum of Understanding (MOU) Between the National Aeronautics and Space Administration and the Ministry of State for Science and Technology [of Canada] for a Cooperative Program Concerning Detailed Definition and Preliminary Design (Phase B) of a Permanently Manned Space Station, April 16, 1985; Memorandum of Understanding Between the United States National Aeronautics and Space Administration and the Science and Technology Agency of Japan for the Cooperative Program Concerning Detailed Definition and Preliminary Design Activities of a Permanently Manned Space Station, May 9, 1985; all in NASA History Office.

CHAPTER 22: MANAGEMENT

1. NASA, Management Colloquium, September 22–23, 1983, NASA History Office.

2. Robert F. Freitag interview conducted by Sylvia D. Fries, July 12, 1985, NASA History Office.

3. NASA Space Station Task Force, "Space Station Management Colloquium White Paper," August 1983, pp. 16, 19, NASA History Office; Jerry W. Craig interview, August 12, 1985.

4. Freitag interview, July 12, 1985.

5. Ibid.

6. NASA, Management Colloquium, September 22–23, 1983; see the three-page sheet labeled "Space Station Management Colloquium, Opening Remarks, Phil Culbertson."

7. Frank Hoban interview, May 28, 1987.

8. John Hodge interview, September 16, 1985; Freitag interview, July 12, 1985.

9. NASA, Management Colloquium, September 22–23, 1983.

10. Statement of General Samuel Phillips, U.S. House Committee on Science and Technology, Subcommittee on Space Science and Applications, August 5, 1986, p. 3, NASA History Office.

11. Neil B. Hutchinson interview, August 14, 1985.

12. "Views of Center Directors on Space Station," in NASA, Management Colloquium, September 22–23, 1983; Frank Hoban, Memorandum for Record, Subject: Space Station Management Colloquium Held at LaRC on September 22–23, 1983, October 17, 1983, attached to NASA, Management Colloquium, September 22–23, 1983.

13. Freitag interview, July 12, 1985. See also Philip Culbertson interview, September 14, 1987; and "Vote taken by Phil Culbertson During the SS Task Force Meeting on Where the Project Office Should Be Located," a one-page sheet, undated, NASA History Office.

14. Hoban interview, May 28, 1987.

15. Ibid.; Hoban, Memorandum for Record, October 17, 1983.

16. James M. Beggs, NASA Administrator, to Gerald D. Griffin, Director, Lyndon B. Johnson Space Center, February 15, 1984; NASA, "Johnson Named Lead NASA Center for Space Station," *NASA News,* release no. 84-25, February 15, 1984; NASA, "Space Station Appointments Announced at Johnson Center," *NASA News,* release no. 84-28, April 9, 1984.

17. Hoban interview, May 28, 1987; Philip Culbertson interview, December 11, 1987.

18. Culbertson interview, December 11, 1987.

19. Ibid.

20. See NASA Space Station Task Force, "Space Station Program Description Document," bk. 4, "Advanced Development Program," chap. 2, pp. 3–4.

21. James M. Romero interview, August 16, 1985; NASA, "NASA Sets Up Space Station Development Teams," *NASA News,* release no. 84-31, February 29, 1984.

22. Culbertson interview, December 11, 1987; Andrew Stofan interview conducted by Adam Gruen, May 7, 1987.

23. Mark B. Nolan interview, August 14, 1985; Freitag interview, July 12, 1985.

24. NASA, "NASA Sets Up Space Station Development Teams." See also U.S. House, Committee on Appropriations, Subcommittee on HUD-Independent Agencies, *Department of Housing and Urban Development–Independent Agencies Appropriations for 1985,* 98th Cong., 2d sess., 1984, pt. 6, pp. 36–38; Culbertson interview, September 14, 1987; and Stofan interview, May 7, 1987.

25. Culbertson interview, December 11, 1987; NASA, "Four NASA Centers Assigned Space Station Studies," *NASA News,* release no. 84-85, June 28, 1984.

26. James M. Beggs, NASA Special Announcement, Subject: Establishment of an Interim Space Station Program Office, April 6, 1984, NASA History Office.

CHAPTER 23: CONGRESS II

1. NASA Office of Space Station, "Space Station: A Man-Tended Approach," report submitted to the Committee on Appropriations, U.S. House of Representatives, and the Committee on Appropriations, U.S. Senate, May 1986.

2. U.S. Congress, *Congressional Record,* 98th Cong., 2d sess., May 30, 1984, p. 14416.

3. Congressman Bill Green, News Release, "Concerns About Proposed Manned Space Station," April 30, 1983 (a typographical error that should read 1984), NASA History Office.

4. Terence Finn interview, June 19, 1986; Hans Mark interview, November 11, 1987; Hans Mark to Jim Beggs, February 26, 1984, NASA History Office.

5. U.S. Congress, Office of Technology Assessment, *Civilian Space Stations*

and the U.S. Future in Space (Washington, D.C.: Government Printing Office, November 1984), p. 4.

6. U.S. House, Committee on Science and Technology, Subcommittee on Space Science and Applications, *1985 NASA Authorization,* 98th Cong., 2d sess., 1984, vol. 2, p. 88; U.S. Senate, Committee on the Budget, *First Concurrent Resolution on the Budget—Fiscal Year 1985,* 98th Cong., 2d sess., 1984, vol. 4, p. 516; U.S. Senate, Committee on Commerce, Science, and Transportation, Subcommittee on Science, Technology, and Space, *NASA Authorization for Fiscal Year 1985,* 98th Cong., 2d sess., 1984, p. 84.

7. Senate Committee on Commerce, Science, and Transportation, *NASA Authorization for Fiscal Year 1985,* pp. 84–85.

8. See House Committee on Science and Technology, *1985 NASA Authorization,* p. 334.

9. Thomas M. Donahue to the Honorable Edward Boland, May 10, 1984; see also James Beggs to the Honorable Edward P. Boland, May 9, 1984, p. 4, NASA History Office.

10. John E. O'Brien, General Counsel, Memorandum to Distribution, Subject: Lobbying, August 19, 1987, NASA History Office.

11. John F. Yardley, President, McDonnell Douglas Astronautics Company, to the Honorable Edward P. Boland, April 30, 1984.

12. Paul J. Burnsky, President, Metal Trades Department, AFL-CIO, to the Honorable Bob Traxler, May 3, 1984.

13. Daniel K. Akaka, Member of Congress, to the Honorable Bill Boner, May 9, 1984. See also Bill Green, Member of Congress, to the Honorable Daniel K. Akaka, May 10, 1984; and Dianne Lambert, Memorandum to Nadia McConnell, Subject: Congressional Groups and Private Organizations Which Can Be Instrumental in Shaping Policy Regarding the Space Station, September 13, 1984. All three documents are on file in the NASA History Office.

14. Hans Mark, *The Space Station: A Personal Journey* (Durham, N.C.: Duke University Press, 1987), p. 207; see also James Beggs interview, October 30, 1987.

15. Beggs to Boland, May 9, 1984.

16. Mark, *Space Station,* p. 207; see also U.S. Congress, *Congressional Record,* 98th Cong., 2d sess., May 30, 1984, pp. 14431, 14434.

17. U.S. House, Committee on Appropriations, Subcommittee on HUD-Independent Agencies, *Department of Housing and Urban Development–Independent Agencies Appropriations for 1985,* 98th Cong., 2d sess., 1984, pt. 6, pp. 39–41.

18. Ibid., pp. 33, 34.

19. Ibid., pp. 36–38.

20. NASA, "Lewis is Assigned Power System Responsibility for Space Station," *NASA News,* release no. 84-46, June 29, 1984; NASA, "Four NASA Centers Assigned Space Station Studies," *NASA News,* release no. 84-85, June 28, 1984; both in NASA History Office.

21. House Committee on Appropriations, *Department of Housing and Urban Development–Independent Agencies Appropriations for 1985,* pp. 32, 34, 35; see also Philip Culbertson interview, September 14, 1987.

22. HUD Committee Print, Report to Accompany H.R. 5713, Department of Housing and Urban Development–Independent Agencies Appropriation Bill, 1985,

June 00 [*sic*], 1984, Draft Only, Subject to Change Until Approved by the Full Committee, p. 66, NASA History Office.

23. U.S. House, Committee on Appropriations, *Department of Housing and Urban Development–Independent Agencies Appropriation Bill, 1985, Committee Report,* 98th Cong., 2d sess., 1984, H. Rpt. 98-803, pp. 35, 65; U.S. Congress, *Congressional Record,* 98th Cong., 2d sess., May 30, 1984, pp. 14416, 14432.

24. James Beggs to the Honorable Edwin (Jake) Garn, June 5, 1984, NASA History Office.

25. Space Station Task Force, file marked "Why We Don't Like HAC Language" (undated [1984]); see also Beggs to Boland, May 9, 1984, p. 2.

26. U.S. Congress, *Congressional Record,* 98th Cong., 2d sess., May 30, 1984, pp. 14431–14433; House Committee on Science and Technology, *1985 NASA Authorization,* vol. 2, pp. 1, 3; Don Fuqua and Larry Winn to the Honorable Edwin Meese, November 21, 1983, NASA History Office.

27. U.S. Congress, *Congressional Record,* 98th Cong., 2d sess., May 30, 1984, p. 14432.

28. Public Law 98-371, An Act Making Appropriations for the Department of Housing and Urban Development, and for Sundry Independent Agencies (98 Stat. 1225), 98th Cong., 2d sess., July 18, 1984. See also U.S. House of Representatives, Conference Report 98-867, [to accompany H.R. 5713], 98th Cong., 2d sess., June 26, 1984, p. 20; M. D. Kerwin to Phil Culbertson et al., Proposed Modified Appropriations Bill and Report Language, June 18, 1984; Culbertson interview, September 14, 1987; and John D. Hodge, "Space Station Presentation to the NASA Advisory Council," July 24, 1985, viewgraph titled "Man-tended Space Station Study."

29. Public Law 98-371, 98 Stat. 1227; NASA Office of Space Station, *Space Station: A Man-tended Approach,* report submitted to the Committee on Appropriations, House of Representatives, and the Committee on Appropriations, U.S. Senate, April 1986; NASA, Advanced Technology Advisory Committee, *Advancing Automation and Robotics Technology for the Space Station and for the U.S. Economy,* vol. 1, *Executive Overview,* report submitted to the U.S. Congress, April 1, 1985; and NASA Office of Space Station, *Automation and Robotics Implementation Activities,* response to the 1985 recommendations of the Advanced Technology Advisory Committee, submitted to the Committee on Commerce, Science, and Transportation, U.S. Senate, March 1986.

30. Administration of Ronald Reagan, "United States Space Policy," *Weekly Compilation of Presidential Documents,* July 4, 1982, pp. 870, 874; Administration of Ronald Reagan, "National Space Strategy: Fact Sheet," reproduced in "National Space Strategy Approved," *Spaceworld,* January 1985, pp. 8–12.

31. James M. Beggs, NASA Special Announcement, Subject: Establishment of the Permanent Office of Space Station, July 27, 1984. See also NASA, "NASA Headquarters Space Station Office Established," *NASA News,* release no. 84-104, July 27, 1984; and Roger Bilstein, "The Space Station—Early Configurations and Phase-B Studies: Narrative and Chronology," prepared for the Johnson Space Center, NASA, contract NAS-17369, October 1986. All three documents are on file in the NASA History Office.

AFTERWORD: POLITICS, BUREAUCRACY, AND PUBLIC POLICY

1. Aaron Wildavsky, *The New Politics of the Budgetary Process* (Glenview, Ill.: Scott, Foresman & Co., 1988).

2. Space Task Group, *The Post-Apollo Space Program: Directions for the Future,* Report to the President (Washington, D.C.: Executive Office of the President, September 1969).

3. John F. Kennedy, Memorandum for the Vice President, April 20, 1961, NASA History Office.

4. Space Station Task Force roundtable interview, August 4, 1987.

5. Charles Lindblom, "The Science of 'Muddling Through,'" *Public Administration Review* 19 (Spring 1959): 79–88.

6. U.S. Senate, Committee on Commerce, Science, and Transportation, Subcommittee on Science, Technology, and Space, *Civil Space Station,* 98th Cong., 2d sess., 1983, p. 53.

7. Space Station Task Force roundtable interview, August 4, 1987.

8. NASA Office of Space Station, "Proceedings of the Space Station Evolution Workshop, Williamsburg, Virginia, September 10–13, 1985," NASA History Office.

9. Senate Committee on Commerce, Science, and Transportation, *Civil Space Station,* p. 48.

10. Donald P. Hearth, "NASA Project Management Study," January 21, 1981, NASA History Office.

11. Space Station Task Force roundtable interview, August 4, 1987.

12. T. O. Paine to Clinton P. Anderson, November 21, 1969, NASA History Office.

13. Space Station Task Force roundtable interview, August 4, 1987.

14. U.S. Senate, Committee on Commerce, Science, and Transportation, Subcommittee on Science, Technology, and Space, *NASA Authorization for Fiscal Year 1984,* 98th Cong., 1st sess., 1983, p. 51.

15. Paine to Anderson, November 21, 1969.

16. "Response to NBC Inquiry on Shuttle and Other Manned Space Flight Program Costs," March 10, 1981; NASA, "Space Shuttle: Fact Sheet," February 1972, appendix entitled "Space Shuttle Economics"; both in NASA History Office.

17. Space Station Task Force roundtable interview, August 4, 1987.

18. James E. Miller, Memorandum for the President, Revised Cost Estimates for the Space Station, February 10, 1987, NASA History Office.

19. NASA, "NASA Proceeding Toward Space Station Development," *NASA News,* release no. 87-50, April 3, 1987.

20. NASA, "News Conference on Space Shuttle," *NASA News,* March 15, 1972, NASA History Office.

21. Courtney G. Brooks, James M. Grimwood, and Loyd S. Swenson, Jr., *Chariots for Apollo: A History of Manned Lunar Spacecraft,* SP-4205 (Washington, D.C.: NASA, 1979), p. 85.

22. Brooks, Grimwood, and Swenson, *Chariots for Apollo,* pp. 190–193.

23. Public Law 100-404, Department of Housing and Urban Development–Independent Agencies Appropriations Act, 1989, 102 Stat. 1026.

24. James Beggs interview, October 30, 1987.

PHOTO CREDITS

Von Braun wheel: *Chesley Bonestell/Bonestell Space Art*
Salyut space station: *TASS/Sovfoto*
Gil Rye: *Reagan Presidential Materials Staff*
Cabinet Council: *Reagan Presidential Materials Staff*
State of the Union: *Bill Fitz-Patrick, The White House*
1987 Space Station: *NASA/Vincent di Fate*
All others: *National Aeronautics and Space Administration*

INDEX